Matrix Inequalities
for
Iterative Systems

T0187496

Hanjo Täubig

Department of Computer Science
Technische Universität München
Garching, Germany

CRC Press
Taylor & Francis Group
Boca Raton London New York

CRC Press is an imprint of the
Taylor & Francis Group, an **informa** business

A SCIENCE PUBLISHERS BOOK

CRC Press
Taylor & Francis Group
6000 Broken Sound Parkway NW, Suite 300
Boca Raton, FL 33487-2742

First issued in paperback 2020

ISBN-13: 978-1-4987-7777-3 (hbk)
ISBN-13: 978-0-367-78260-3 (pbk)

Library of Congress Cataloging-in-Publication Data

Names: Täubig, Hanjo, 1975- | Hou, Xu (Engineer), editor.
Title: Matrix inequalities for iterative systems / Hanjo Täubig, Department of Computer Science, Technische Universität München, Garching, Germany, Design, Fabrication, Properties, and Applications of Smart and Advanced Materials ; editor, Xu Hou, Harvard University, School of Engineering and Applied Sciences.
Description: Boca Raton, FL : CRC Press, 2017. | "A science publishers book." | Includes bibliographical references and index.
Identifiers: LCCN 2016030679| ISBN 9781498777773 (hardback) | ISBN 9781498777797 (e-book)
Subjects: LCSH: Matrix inequalities.
Classification: LCC QA188 .T36 2017 | DDC 512.9/434--dc23
LC record available at https://lccn.loc.gov/2016030679

Visit the Taylor & Francis Web site at
http://www.taylorandfrancis.com

and the CRC Press Web site at
http://www.crcpress.com

Matrix Inequalities
for
Iterative Systems

To my parents
Brigitte and Klaus Täubig

Acknowledgments

It seems to me shallow and arrogant for any man in these times to claim he is completely self-made, that he owes all his success to his own unaided efforts. Many hands and hearts and minds generally contribute to anyone's notable achievements.

Walt Disney

First of all, I want to thank Prof. Ernst Mayr for the generous support through all the years and for providing me with the freedom to work on whatever seemed to be interesting to me. I am also deeply indebted to all the other people that contributed to the work on the number of walks, especially the members of our "NetLEA" group (consisting of Sven Kosub, Klaus Holzapfel, Moritz Maaß, Alexander Offtermatt-Souza, and me). When we started in spring of 2002 to work on spectral graph theory, we made an attempt to prove our first conjectures. Some of these early ideas are still at the heart of the proofs for the much more general statements that we showed only recently. Numerous important ideas, in particular for the only long and complex proof in this work, were contributed by Jeremias Weihmann. I am thankful for his contributions and for all the fruitful discussions that we had. I am also grateful for all the hints that I got from Raymond Hemmecke, Thomas Kahle, David Reeb, Werner Meixner, and from my brother Holger Täubig. For quite recent interesting remarks and discussions, I am thankful to Tamás Réti. For proofreading a draft of this work, I am indebted to Moritz Fuchs.

It was a great pleasure to meet the real experts at the Conference on Applications of Graph Spectra in Computer Science (Barcelona, 2012) and at the Conference on Algebraic Combinatorics: Spectral Graph Theory, Erdős-Ko-Rado Theorems and Quantum Information Theory, a conference to celebrate the work of Chris Godsil (Waterloo, 2014).

Furthermore, I want to thank all the other people that I had the pleasure to work with, in particular, all the members of the efficient algorithms group at TUM (especially Riko Jacob, Christian Scheideler, Stefan Schmid, and Harald Räcke).

Kind regards go to the members of my ProLehre teaching course and to our coaches Barbara Greese, Hans-Christoph Bartscherer, Pit Forster, and Adi Winteler.

Preface

Our interest for the topic was attracted by a conjecture that was brought up by our colleague Sven Kosub. While we did not make much progress in proving the conjecture at the beginning (because it was wrong), it turned out later, that one of the first ideas for proving a certain special case was still at the heart of the more general results that were obtained only recently by Jeremias Weihmann and the author of this book.

This book is based on the author's habilitation thesis [Täu15a]. The main goal of the habilitation process is to develop and prove the ability to teach. As such, the aim of the book is to present the topic in such a way that the results and their underlying methods of proof can be used by the reader in the easiest possible way. To this end, our work tries to unify the various inequalities for the number of walks in graphs and for the sum of entries of matrix powers in a generalized form. These generalizations reveal the fundamental principles underlying the different results.

To the best of our knowledge all the results claimed to be new, in this work, have never appeared before. During the whole time of working on this subject, we found many examples of results that were (re)discovered more than just once. But in view of the large list of references, we can probably claim that we did a thorough job for finding most of the relevant literature. In any case, the focus of this book is not only on the collection, but also on the systematization of the vast number of results. We have generalized and unified the related results, and put them into a common frame and discussed their relations. In conclusion, we hope to contribute to a deeper understanding of the origin of the different inequalities for matrix powers and the number of walks in graphs. Our main goal is thus a clear presentation of the common principles underlying the tremendous number of different results.

Almost all of the proofs in this book are elementary. While it is normal to be satisfied with an arbitrary complex proof for a certain statement, it must be emphasized that elementary proofs (even for known results) have an additional value because they are instructive and convincing.

Most of the results in this book have been published in the following articles and reports: [HKM+11; HKM+12; Täu12; TW12; TWK+13; TW14; Täu14; Täu15b].

Contents

Symbol Description

A	matrix with entries a_{ij}, usually square of size $n \times n$	G	graph, usually with vertex set V and edge set E
$A[X, Y]$	submatrix of A induced by row index set X and column index set Y	V	vertex set of a graph
$A[X]$	principal submatrix of A induced by row and column index set X	E	edge set of a graph
A^T	transpose of A	$d_{in}(v)$	in-degree of vertex v
A^*	conjugate transpose of A	$d_{out}(v)$	out-degree of vertex v
a_{ij}	matrix entry that resides in row i and column j	Δ	maximum degree
I_n	$n \times n$ identity matrix	$\rho(G)$	edge density of the graph G
$\text{sum}(A)$	sum of all entries of A	$w_k(x, y)$	number of walks of length k starting at vertex x and ending at vertex y
$r_i(A), r_i$	row sum of the entries in row i of the given matrix A	$s_k(x)$	number of walks of length k starting at vertex x
$c_j(A), c_j$	column sum of the entries in column j of the given matrix A	$e_k(x)$	number of walks of length k ending at vertex x
$\text{tr}(A)$	trace of matrix A (sum of main diagonal entries)	$w_k(x)$	replacement for $s_k(x)$ and $e_k(x)$ in undirected graphs
$\rho(A)$	spectral radius of A	w_k	total number of walks of length k in a given graph
$\langle x, y \rangle$	(standard) inner product of vectors x and y	cl_k	total number of closed walks of length k in a given graph
$\|x\|$	(Euclidean) length of vector x	$cl_k(x)$	number of closed walks of length k starting at vertex x
1_n	n-dimensional all-ones vector	v_k	total number of nonreturning walks of length k in a given graph
$\chi(S)$	characteristic vector of set S		

INTRODUCTION

I

Chapter 1

Notation and Basic Facts

There is no problem in all mathematics that cannot be solved by direct counting.
But with the present implements of mathematics many operations can be performed
in a few minutes which without mathematical methods would take a lifetime.

Ernst Mach

We use standard notation, where $\mathbb{N}$ denotes the set of nonnegative integers, $\mathbb{R}$ is the set of real numbers, $\mathbb{R}_{\geq 0}$ denotes the set of nonnegative real numbers, and $\mathbb{C}$ is the set of complex numbers. For convenience, the set of integers $\{1, \ldots, n\}$ is abbreviated by $[n]$. Frequently used notation can be found in the symbol description at the beginning of the book.

Suppose that $c \in \mathbb{C}$ is a complex number of the form $c = a + ib$ with $a, b \in \mathbb{R}$. The *real part* of c is denoted by $\Re(c) := a$ and the *complex conjugate* of c is denoted by $\bar{c} := a - ib$. The sum $c + \bar{c} = a + ib + a - ib = 2a = 2\Re(c)$ is a real number. The product $c\bar{c} = (a + ib)(a - ib) = a^2 + b^2$ is a nonnegative real number and its square root defines the *modulus* or *absolute value* of c, which is denoted by $|c| := \sqrt{a^2 + b^2} = \sqrt{c\bar{c}}$.

1.1 Matrices and Vectors

1.1.1 Matrices

Throughout this book, A denotes a matrix with complex entries $a_{i,j} := A_{(i,j)} \in \mathbb{C}$, where i refers to the row and j refers to the column of A. Unless stated otherwise, A is assumed to be a square $n \times n$-matrix. In some cases, we will discuss rectangular matrices where the term $m \times n$-matrix refers to a matrix having m rows and n columns. We refer to the entry in row i and column j of a matrix A (i.e.,

$A_{(i,j)} = a_{i,j}$) as the (i,j)-entry of A. The *transpose* A^T of an $m \times n$-matrix A is the $n \times m$-matrix defined by $(A^T)_{(i,j)} = A_{(j,i)}$. Let $\bar{A}$ with $\bar{A}_{(i,j)} = \overline{A_{(i,j)}}$ denote the component-wise complex conjugate of A. Then, the *conjugate transpose*[1] A^* of a matrix A is defined by $A^* := \bar{A}^T$. The *identity matrix*, i.e., the square $n \times n$-matrix having all main diagonal entries a_{ii} equal to one, and all other entries a_{ij} $(i \neq j)$ equal to zero, is denoted by I_n. If the dimension is clear from the context, we leave out the index n. Given a square $n \times n$-matrix A, a matrix B satisfying $AB = I_n$ is called the *inverse matrix* of A. If A has an inverse, then A is called *invertible* (or *nonsingular*). The inverse of an invertible matrix A is unique and denoted by A^{-1}.

For a given $m \times n$-matrix A, a subset $X \subseteq [m]$ of row indices, and a subset $Y \subseteq [n]$ of column indices, the *submatrix* of all elements lying in a row indexed by X and in a column indexed by Y is denoted by $A[X,Y]$. If A is a square matrix $(m = n)$, then the square submatrix $A[X,X]$ is called a *principal submatrix* and it is abbreviated by $A[X]$.

1.1.2   Vectors

A column vector $x \in \mathbb{C}^n$ consisting of n entries $x_i \in \mathbb{C}$ is a rectangular matrix having n rows and just one column. Accordingly, a row vector is a rectangular $1 \times n$-matrix. Hence, the matrix operations can also be applied to vectors to define x^T as the corresponding row vector and x^* as the conjugate transpose of x. A concept that generalizes the usual scalar or dot product $\langle x, y \rangle = x^T y$ of real vectors $x, y \in \mathbb{R}^n$ is the *standard inner product* for complex vectors $x, y \in \mathbb{C}^n$. It is defined as

$$\langle x, y \rangle := x^* y \ .$$

Note that different conventions are used for defining which of the two arguments is linear and which is conjugate-linear (or antilinear). Here, we use conjugate linearity in the first (x) and linearity in the second (y) argument. That means, we have $\langle \alpha x_1 + \beta x_2, y \rangle = \bar{\alpha} \langle x_1, y \rangle + \bar{\beta} \langle x_2, y \rangle$ and $\langle x, \alpha y_1 + \beta y_2 \rangle = \alpha \langle x, y_1 \rangle + \beta \langle x, y_2 \rangle$ for all vectors $x, y, x_1, x_2, y_1, y_2 \in \mathbb{C}^n$ and numbers $\alpha, \beta \in \mathbb{C}$. Two vectors $x, y \in \mathbb{C}^n$ are called *orthogonal* if and only if $\langle x, y \rangle = 0$. Associated with each vector x is its (Euclidean) *length* $\|x\| := \sqrt{\langle x, x \rangle}$. A vector of length 1 is called a *unit vector*. For every vector x with $\|x\| \neq 0$, the *normalized vector* $x/\|x\|$ is a unit vector pointing into the same direction as x.

In the context of submatrices, we will frequently use a *characteristic vector*[2] $\chi(S)$ for a certain subset S of a fixed basic set U. Each entry $\chi_i(S)$ for $i \in U$ indicates whether $i \in S$ or $i \notin S$ by the definition

$$\chi_i(S) = \begin{cases} 1 & \text{if } i \in S, \\ 0 & \text{if } i \notin S. \end{cases}$$

The universe U will be clear from the context. It will be either the vertex set V of a multigraph or the row (or column) index set $[m]$ (or $[n]$) of an $m \times n$-matrix.

---

[1]also known as Hermitian transpose/conjugate/adjoint.
[2]also known as *indicator vector*.

One particular special case of those vectors is the n-dimensional all-ones vector $\mathbf{1}_n := (\underbrace{1, \ldots, 1}_{n\text{-times}})^T$.

1.1.3 Matrix Classes

In many places, we will restrict the basic set of matrices that we consider to be one of the following subclasses of *complex* matrices (matrices whose entries are arbitrary complex numbers). A matrix A is said to be *real* if all its entries are real numbers. Such a matrix is called *nonnegative* if all of its entries are nonnegative real numbers. A particularly interesting subclass of those matrices are the *stochastic* matrices. Depending on the context, they are either required to have all row sums or all column sums equal to 1. Mostly they are used to describe the behavior of Markov chains, i.e., the transitions in certain probabilistic processes. Another important subclass of the nonnegative matrices is the class of $\{0,1\}$-*matrices*. Here, the basic set of the entries is restricted by $a_{ij} \in \{0,1\}$. They can be used to describe the edge structure of directed or undirected graphs.

A square $n \times n$-matrix A is *symmetric* if and only if $a_{ij} = a_{ji}$ for all $i, j \in [n]$, i.e., A is equal to its transpose ($A = A^T$). A subclass of the symmetric matrices is the class of *diagonal* matrices where all entries residing outside the main diagonal are required to be zero ($a_{ij} = 0$ for $i \neq j$). A matrix A is *self-adjoint* or *Hermitian* if and only if it is equal to its conjugate transpose ($A = A^*$). In the case where $A^* = -A$, the matrix A is called *skew-Hermitian*. A matrix A is *normal* if and only if $A^*A = AA^*$, i.e., if A commutes with its conjugate transpose. A complex square $n \times n$-matrix is *unitary* if and only if $A^*A = AA^* = I_n$. A unitary real matrix is called *orthogonal*. Since the defining condition reduces to $A^TA = AA^T = I$ in this case, there is another possible characterization that requires the transpose to be equal to the inverse ($A^T = A^{-1}$).

A matrix A is *positive-semidefinite* if and only if $x^*Ax \geq 0$ for all $x \in \mathbb{C}^n$. This is equivalent to the condition that A is Hermitian and has only nonnegative eigenvalues. Accordingly, A is *negative-semidefinite* if and only if $x^*Ax \leq 0$ for all $x \in \mathbb{C}^n$. This is equivalent to the condition that A is Hermitian and all its eigenvalues are nonpositive. A further equivalent definition would be that $-A$ is positive-semidefinite. Note that normal matrices contain the classes of orthogonal, unitary, real symmetric, real skew-symmetric, positive/negative-semidefinite, Hermitian and skew-Hermitian matrices.

1.2 Graphs

Throughout this work, $G = (V, E)$ denotes a finite (directed or undirected) multigraph with a set V of n vertices and a multiset E of m edges. An *undirected edge* between vertices u and v is denoted by the unordered pair $\{u, v\}$. In contrast, a *directed edge* from u to v is denoted by the ordered pair (u, v). An edge of the form $\{u, u\}$ or (u, u) that connects a vertex u to itself is called a *loop*. Unless stated otherwise, we assume that the term *graph* refers to a *simple graph* that is not allowed to contain parallel edges or loops.

1.2.1 Degrees

The *out-degree* of a vertex x (the number of edges emanating from x) is denoted by $d_{\mathrm{out}}(x)$ while the *in-degree* of x (the number of edges pointing to x) is denoted by $d_{\mathrm{in}}(x)$. For undirected graphs, there is no need to distinguish between in-degree and out-degree. In this case, the *degree* is denoted by $d(x)$. The same applies to directed graphs having $d_{\mathrm{out}}(x) = d_{\mathrm{in}}(x)$, which coincides with the set of graphs that are disjoint unions of graphs having an Eulerian cycle. For $n \geq 1$, the *average degree* is denoted by

$$\bar{d} := \frac{1}{n} \sum_{v \in V} d_{\mathrm{in}}(v) = \frac{1}{n} \sum_{v \in V} d_{\mathrm{out}}(v) \ .$$

For directed graphs, we have $\sum_{v \in V} d_{\mathrm{in}}(v) = \sum_{v \in V} d_{\mathrm{out}}(v) = m$, that is $\bar{d} = \frac{m}{n}$. The situation is slightly different for undirected graphs. Here, every edge contributes 1 to the degree of both incident vertices, and we have $\sum_{v \in V} d(v) = 2m$ for graphs without loops (which is sometimes called the *handshake lemma*). This implies $\bar{d} = \frac{2m}{n}$.

1.2.2 Adjacency Matrix

The *adjacency matrix* A of a (directed or undirected) multigraph G on the vertex set $\{v_1, \ldots, v_n\}$ is the $n \times n$-matrix where, for all $i, j \in [n]$, each entry $a_{i,j}$ is equal to the *number of edges* from vertex v_i to vertex v_j in G. In simple graphs, this corresponds to

$$a_{i,j} = \begin{cases} 1 & \text{if there is an edge from } i \text{ to } j \text{ in } G, \\ 0 & \text{otherwise.} \end{cases}$$

Most of the results are valid for the even more general concept of edge-weighted graphs where the entries $a_{i,j}$ of the weighted adjacency matrix A would be defined to be the weight of the edge(s) going from i to j.

1.3 Number of Walks in Graphs

1.3.1 Walks

A *walk*[3] in a multigraph $G = (V, E)$ is an alternating sequence

$$(v_0, e_1, v_1, \ldots, v_{k-1}, e_k, v_k)$$

of vertices $v_i \in V$ and edges $e_i \in E$ (that starts and ends with a vertex) where each edge e_i of the walk must connect vertex v_{i-1} to vertex v_i in G, that is, $e_i = (v_{i-1}, v_i)$ is required for directed graphs and $e_i = \{v_{i-1}, v_i\}$ is required for undirected graphs.

---

[3]In some publications, this is called a *path*, see, e.g., Flajolet and Sedgewick [FS09].

Vertices and edges can be used repeatedly in the same walk. If the multigraph has no parallel edges, the walks could also be specified by the sequence of vertices without the edges. The *length* of a walk is the number of edge traversals. That means, the walk $(v_0, \ldots, v_k)$ consisting of $k+1$ vertices and k edges is a walk of length k. Mostly, we will call it a *k-step walk*.[4]

A walk that starts and ends at the same vertex is called a *closed walk*. Sometimes, another special class of walks in undirected graphs is considered. A *nonreturning walk* in an undirected graph G is a walk $(v_0, e_1, v_1, e_2, \ldots, v_{k-1}, e_k, v_k)$ such that $e_i \neq e_{i+1}$ for $1 \leq i \leq k-1$, i.e., it does never traverse the same edge in any two consecutive steps.

1.3.2 Number of Walks

Our main concern will be the investigation of the *number of walks* of a specified length. For vertices $x, y \in V$ and $k \in \mathbb{N}$, let $W_k(x, y)$ denote the set of walks of length k that start at vertex x and end at vertex y. Let

$$w_k(x, y) := |W_k(x, y)|$$

denote the corresponding number of walks of length k from x to y. For undirected graphs, we have $w_k(x, y) = w_k(y, x)$. We extend this notation to sets in the following way: For vertex subsets $X, Y \subseteq V$, the set of walks of length k starting at a vertex of X and ending at a vertex of Y is denoted by $W_k(X, Y)$. The corresponding number of those walks is denoted by $w_k(X, Y) := |W_k(X, Y)|$.

Furthermore, let

$$s_k(x) := w_k(\{x\}, V) = \sum_{y \in V} w_k(x, y) \qquad \text{and} \qquad e_k(x) := w_k(V, \{x\}) = \sum_{y \in V} w_k(y, x)$$

denote the number of k-step walks that *start* at vertex x and the number of k-step walks that *end* at vertex x, respectively. In the case of undirected graphs, those numbers are equal. Then we use the term $w_k(x) := s_k(x) = e_k(x)$.

In general,

$$w_k := \sum_{x \in V} s_k(x) = \sum_{x \in V} e_k(x)$$

denotes the total number of walks of length k. Obviously, $w_k = \sum_{x \in V} w_k(x)$ holds for undirected graphs. The number of closed k-step walks starting and ending at vertex x is denoted by $cl_k(x) := w_k(x, x)$. The total number of closed walks of length k is denoted by

$$cl_k := \sum_{x \in V} cl_k(x) = \sum_{x \in V} w_k(x, x) \ .$$

The number of nonreturning walks of length k is denoted by ν_k.

---

[4] Also the term *k-walk* is used, but we will try to avoid it since other authors use it with a different meaning which generalizes the concept of Hamilton cycles.

Note that, since walks are directed, a walk is in general different from its reversed walk (except for walks with palindromic sequence of vertices). Also, closed walks are in general different from their cyclic shifts. It is a natural approach to extend the concept of walks to *weighted walks*. In this case, the weight of a walk corresponds to the product of the respective edge weights. Such a definition was used, for instance, by Kwapisz [Kwa96], Butler [But06a], and in the book of Brualdi and Cvetković [BC09].

A whole chapter of the book by Stanley [Sta13] is dedicated to walks in undirected multigraphs.

1.4 Entry Sums and Matrix Powers

1.4.1 *Entry Sums of Matrix Powers*

For any $m \times n$-matrix A, the vector of *row sums* and the vector of *column sums* are denoted by $r(A) := A\mathbf{1}_n$ and $c(A) := A^T\mathbf{1}_m$, respectively. Thus, the sum of all entries in row i of A is denoted by $r_i(A)$ and the sum of all entries in column j of A is $c_j(A)$. In most instances, the matrix A is clear from the context. Hence we abbreviate the notation by r_i and r_j. Accordingly, the row and column sum vectors are often abbreviated by r and c. A matrix A is called *sum-symmetric*, if $r_i(A) = c_i(A)$ for all $i \in [n]$. An important subclass of the nonnegative sum-symmetric matrices are the *doubly-stochastic* matrices, where all entries must be nonnegative and all row sums and all column sums are required to be equal to 1.

For the total sum of all entries of A, we use the notation

$$\mathrm{sum}(A) := \sum_{i \in [m]} \sum_{j \in [n]} a_{i,j} = \mathbf{1}_m^T A \mathbf{1}_n \ .$$

For convenience, the (i,j)-entry of the matrix power A^k of a square $n \times n$-matrix A is denoted by

$$a_{i,j}^{[k]} := \left(A^k \right)_{(i,j)} \qquad \text{for } i, j \in [n] \text{ and } k \in \mathbb{N}.$$

Similarly, we define $r^{[k]} := r(A^k)$ and $c^{[k]} := c(A^k)$ as abbreviations for the row and column sums of A^k when A is clear from the context.

For the special case of $k = 0$, we assume that A^k yields the identity matrix $I = A^0$.

1.4.2 *Number of Walks and Matrix Powers*

Regarding walks in a graph and the powers of its adjacency matrix, the following is a well known fact. Here, we assume that $V = \{v_1, \ldots, v_n\}$ and each vertex v_i is associated with row and column index i in A for all $i \in [n]$.

Lemma 1.1
For every multigraph on n vertices with adjacency matrix A, the (i,j)-entry of A^k equals the number of walks of length k that start at vertex v_i and end at vertex v_j, i.e.,

$$w_k(v_i, v_j) = \left(A^k\right)_{(i,j)} = a_{i,j}^{[k]}$$

for $i, j \in [n]$ and $k \in \mathbb{N}$.

Proof   The statement is true for $k = 0$ since $A^0 = I$ and there is exactly one walk of length 0 from each vertex to itself (and there is no other 0-step walk).
It is also valid for $k = 1$ since $A^1 = A$ and each directed edge corresponds to a walk of length 1. Therefore there are exactly $a_{i,j}$ different walks from vertex v_i to vertex v_j.
Now proceed by induction. Assume that the statement is true up to a certain $k \in \mathbb{N}$. The number of walks of length $k + 1$ from v_i to v_j can be calculated by distinguishing the penultimate vertex $v_{j'}$. For all different vertices $v_{j'}$, all walks of length k from v_i to $v_{j'}$ can be combined with all different walks of length 1 (edges) from $v_{j'}$ to v_j. Thus, we obtain $w_{k+1}(v_i, v_j) = \sum_{v_{j'} \in V} w_k(v_i, v_{j'}) \cdot w_1(v_{j'}, v_j)$. By the induction assumption, this is equal to $\sum_{j' \in [n]} a_{i,j'}^{[k]} \cdot a_{j',j}$. This, in turn, is equal to $a_{i,j}^{[k+1]}$ by the definition of matrix multiplication.

Lemma 1.1 implies that the number of k-step walks starting at a certain vertex v_i (or ending at vertex v_j) is the i-th row sum (or j-th column sum, respectively) of the k-th power of the adjacency matrix A, i.e., $s_k(v_i) = r_i^{[k]}$ (or $e_k(v_j) = c_j^{[k]}$). In particular, we observe the following special cases. For walks of length 0, we have $s_0(v) = e_0(v) = 1$ (that is $w_0(v) = 1$ for undirected graphs) for each vertex v. This implies $w_0 = n$ for the total number of 0-step walks. For walks of length 1, we have $s_1(v_i) = d_{\text{out}}(v_i) = r_i$ as well as $e_1(v_i) = d_{\text{in}}(v_i) = c_i$ (that is $w_1(v_i) = d(v_i) = r_i = c_i$ for undirected graphs). This implies $w_1 = \sum_{v \in V} d_{\text{out}}(v) = \sum_{v \in V} d_{\text{in}}(v) = m$ for directed graphs and $w_1 = \sum_{v \in V} d(v) = 2m$ for undirected graphs (by the handshake lemma). The total number of closed walks of length k equals the *trace* (the sum of the entries on the *main diagonal*) of A^k: $cl_k = \sum_{i \in [n]} a_{i,i}^{[k]} = \text{tr}(A^k)$.

1.4.3   Local Decomposition

A fundamental fact is that matrix powers and the number of walks obey the following local decomposition law.[5]

Lemma 1.2 (Local Decomposition)
For every matrix A and $x, y, z \in [n]$, we have

$$a_{x,z}^{[k+\ell]} = \sum_{y \in V} a_{x,y}^{[k]} \cdot a_{y,z}^{[\ell]} \ .$$

---

[5] For Markov chains, results corresponding to Lemma 1.2 are called Chapman–Kolmogorov equations.

Proof   The result follows directly from the definition of matrix multiplication applied to the product $A^k A^\ell = A^{k+\ell}$ (since matrix multiplication is associative).

This means that, in any directed or undirected multigraph $G = (V, E)$, the number of walks of length $k + \ell$ from a vertex $x \in V$ to a vertex $z \in V$ can be decomposed via intermediate vertices y by

$$w_{k+\ell}(x, z) = \sum_{y \in V} w_k(x, y) \cdot w_\ell(y, z) \ .$$

An extensive treatment of walks, matrix powers, and related results can also be found in the book of Brualdi and Cvetković [BC09, p. 57].

1.5   Subsets, Submatrices, and Weighted Entry Sums

1.5.1   Selected Walks and Submatrices

Recall that, for an arbitrary $m \times n$-matrix A, the entry sum of A can be calculated by $\mathbf{1}_m^T A^k \mathbf{1}_n$. Using the m-dimensional characteristic vector $\chi(X)$ and the n-dimensional characteristic vector $\chi(Y)$, the similar formula

$$\mathrm{sum}(A[X, Y]) = \chi(X)^T A \chi(Y)$$

calculates the entry sum for the submatrix $A[X, Y]$ that is induced by the row index set $X \subseteq [m]$ and the column index set $Y \subseteq [n]$. We will mainly apply this to powers A^k of a square $n \times n$-matrix A, in particular to adjacency matrices of graphs.

By applying this scheme together with Lemma 1.1, the total number of k-step walks in a graph $G = (V, E)$ with adjacency matrix A can be counted by $w_k = \mathbf{1}_n^T A^k \mathbf{1}_n$. For counting only the k-step walks starting at a certain vertex set $X \subseteq V$ and ending at a vertex from another set $Y \subseteq V$ (while allowing arbitrary vertices at the inner part of the walk), we can use the formula

$$w_k(X, Y) = \chi(X)^T A^k \chi(Y) \ .$$

In the majority of cases, we will apply this counting scheme to matrix powers A^k for equal sets $X = Y$. Note that the entry sums of $(A^k)[X]$ and $(A[X])^k$ are *not* the same. The first one is a principal submatrix of the matrix power. For an adjacency matrix A, it counts the walks of length k that start and end at a vertex of the set X. Arbitrary vertices are allowed at the inner parts of the walks. In contrast, the second one is the power of the principal submatrix. For an adjacency matrix A, it counts only the walks in the subgraph induced by X, i.e., *all* vertices of the walk have to be in X.

1.5.2   Weighted Entry Sums

More generally, we could sum up the entries of any *real* matrix A (or matrix power A^k) in a weighted form where the entries of each row i are scaled by a certain *real* value x_i and, additionally, each column j is scaled by y_j:

$$\text{sum}_{x,y}\left(A^k\right) := x^T A^k y = \langle x, A^k y \rangle = \langle (x^T A^k)^T, y \rangle \in \mathbb{R} \qquad (x, y \in \mathbb{R}^n).$$

Consider this expression as a function f of the vectors x and y, parameterized by the matrix A. Since the entry in row i and column j is effectively scaled by $x_i y_j$, i.e., $f_{A^k}(x, y) = \text{sum}_{x,y}\left(A^k\right) = \sum_{i=1}^n \sum_{j=1}^n x_i a_{i,j}^{[k]} y_j$, this is called a *bilinear form*. This weighting scheme is also a generalization of the sum of weighted walks derived from a matrix A as defined by Brualdi and Cvetković [BC09, p. 51]. If A is the identity matrix I then the bilinear form reduces to the scalar product of the real vectors x and y.

The weighting scheme can be generalized even further by applying the same principle to *complex* matrices A and *complex* weight vectors x and y:

$$\text{sum}_{x,y}\left(A^k\right) := x^* A^k y = \langle x, A^k y \rangle = \langle (x^* A^k)^*, y \rangle \in \mathbb{C} \qquad (x, y \in \mathbb{C}^n).$$

This corresponds to $f_A(x, y) := \text{sum}_{x,y}\left(A^k\right) = \sum_{i=1}^n \bar{x}_i a_{i,j}^{[k]} y_j$. Since this function is linear in y and conjugate-linear in x, it is called a *sesquilinear form*. If A is a Hermitian matrix then $f_A(x, y) = \overline{f_A(y, x)}$. A sesquilinear form that obeys this condition is called a *Hermitian form* (or symmetric sesquilinear form). For the special case $A = I$, this Hermitian form reduces to the standard inner product $\sum_{i=1}^n \bar{x}_i y_i = \langle x, y \rangle$.

1.5.3 Quadratic Forms and Hermitian Matrices

If A is an $n \times n$ Hermitian matrix then the sum of all entries is a *real* number ($\mathbf{1}_n^T A \mathbf{1}_n \in \mathbb{R}$). Also the sum of all entries for any *principal submatrix* induced by an index set $S \subseteq [n]$ is a real number ($\chi(S)^T A \chi(S) \in \mathbb{R}$). In particular, this applies to each entry on the main diagonal.

By multiplying each row i and column i of a Hermitian matrix A with the *same* respective scaling factor $x_i = y_i \in \mathbb{R}$ (i.e., if $x = y \in \mathbb{R}^n$), we obtain a Hermitian matrix again. Thus, using the *quadratic form*

$$\text{sum}_s(A) := \text{sum}_{s,s}(A) = s^T A s = \langle s, A s \rangle \in \mathbb{R} \qquad (s \in \mathbb{R}^n) \ ,$$

the weighted sum of all entries is again a real number. Of course, the same applies to the powers of the matrix:

$$\text{sum}_s\left(A^k\right) = \text{sum}_{s,s}\left(A^k\right) = s^T A^k s = \langle s, A^k s \rangle \in \mathbb{R} \qquad (s \in \mathbb{R}^n) \ .$$

Note that the quadratic form $q_A(x) := h_A(x, x)$ for any Hermitian form $h_A(x, y) = \overline{h_A(y, x)}$ is always a *real* number. Thus, for a Hermitian matrix A, we have

$$\text{sum}_s(A) = s^* A s = \langle s, A s \rangle \in \mathbb{R} \qquad (s \in \mathbb{C}^n) \ .$$

In particular, we are interested in the weighted sums of powers of a Hermitian matrix A:

$$\text{sum}_s\left(A^k\right) = s^* A^k s = \langle s, A^k s \rangle \in \mathbb{R} \qquad (s \in \mathbb{C}^n) \ .$$

For quadratic forms, we also define the following weighted versions of the matrix power entries:

$$a_{i,j}^{[k,s]} := \bar{s}_i s_j \left(A^k\right)_{(i,j)} = \bar{s}_i s_j a_{i,j}^{[k]} \; .$$

For indices $i, j \in [n]$, let $r_i^{[k,s]}$ and $c_j^{[k,s]}$ denote the *weighted row and column sums* of the terms $a_{i,j}^{[k]}$:

$$r_i^{[k,s]} = \sum_{j=1}^{n} a_{i,j}^{[k,s]} \qquad \text{and} \qquad c_j^{[k,s]} = \sum_{i=1}^{n} a_{i,j}^{[k,s]} \; .$$

For further information on quadratic and sesquilinear (Hermitian) forms, we refer to the respective chapter in Gantmacher [Gan60].

1.5.4 Global Decomposition

Similar to the local decomposition in Lemma 1.2, there is a global decomposition for the entry sum of matrix powers.

Lemma 1.3 (Global 3-Parts Decomposition)
For every $n \times n$ matrix A, we have

$$\text{sum}\left(A^{\ell_1 + \ell_2 + \ell_3}\right) = \sum_{i \in [n]} \sum_{j \in [n]} c_i^{[\ell_1]} \cdot a_{i,j}^{[\ell_2]} \cdot r_j^{[\ell_3]} \; .$$

Proof   Since matrix multiplication is associative, we have

$$\text{sum}\left(A^{\ell_1 + \ell_2 + \ell_3}\right) = \mathbf{1}_n^T \left(A^{\ell_1} A^{\ell_2} A^{\ell_3}\right) \mathbf{1} = \left(\mathbf{1}_n^T A^{\ell_1}\right) A^{\ell_2} \left(A^{\ell_3} \mathbf{1}\right) = c^{[\ell_1]^T} A^{\ell_2} r^{[\ell_3]} \; .$$

Note that a similar result can be obtained for the weighted case:

$$\text{sum}_{x,y}\left(A^{\ell_1 + \ell_2 + \ell_3}\right) = x^* \left(A^{\ell_1} A^{\ell_2} A^{\ell_3}\right) y = \left(x^* A^{\ell_1}\right) A^{\ell_2} \left(A^{\ell_3} y\right) \; .$$

Similar to the calculation of the total number of ℓ_2-step walks by the formula $\mathbf{1}_n^T A^{\ell_2} \mathbf{1}_n$, Lemma 1.3 means that we can calculate the total number of $(\ell_1 + \ell_2 + \ell_3)$-step walks by multiplying the matrix of ℓ_2-walks with the vector e_{ℓ_1} of ℓ_1-walks that end at each vertex and the vector s_{ℓ_3} of ℓ_3-walks that start at each vertex. Thus, using the formula $e_{\ell_1} A^{\ell_2} s_{\ell_3}$, each of the ℓ_2-walks from vertex x to vertex y is multiplied by $e_{\ell_1}(x)$ and $s_{\ell_3}(y)$. For the special case of undirected graphs, setting $\ell_1 = \ell_3 = \ell$ and $\ell_2 = k$ leads to counting the walks of length $k + 2\ell$ using a quadratic form.

An important special case of Lemma 1.3 occurs when the middle part is empty, i.e., $\ell_2 = 0$ and therefore $A^{\ell_2} = A^0 = I$ is the identity matrix. Then we have $a_{i,j}^{[0]} = 1$ if $i = j$; otherwise $a_{i,j}^{[0]} = 0$.

Corollary 1.1 (Global 2-Parts Decomposition)
For every $n \times n$ matrix A, we have

$$\text{sum}\left(A^{k+\ell}\right) = \sum_{i \in [n]} c_i^{[k]} \cdot r_i^{[\ell]} \ .$$

Using weighting vectors x and y, this could be generalized to

$$\text{sum}_{x,y}\left(A^{k+\ell}\right) = \left(x^* A^k\right)\left(A^\ell y\right) \ .$$

For walks in directed and undirected graphs, the corresponding equalities are

$$w_{k+\ell} = \sum_{v \in V} e_k(v) \cdot s_\ell(v) \qquad \text{and} \qquad w_{k+\ell} = \sum_{v \in V} w_k(v) \cdot w_\ell(v) \ .$$

As a special case, the decomposition lemma implies

$$w_{k+1} = \sum_{v \in V} d_{\text{in}}(v) \cdot s_k(v) = \sum_{v \in V} e_k(v) \cdot d_{\text{out}}(v) \ .$$

for directed graphs. For undirected graphs, this translates to

$$w_{k+1} = \sum_{v \in V} d(v) \cdot w_k(v) = \sum_{\{v,w\} \in E} w_k(v) + w_k(w) \ .$$

1.6 Elementary Inequalities for Vectors and Sequences

1.6.1 The Inequalities of Cauchy, Bunyakovsky, and Schwarz

One of the most fundamental statements for inner product spaces is the Cauchy-Schwarz inequality. Cauchy [Cau21, Thm. 16, p. 455] proved the following basic form.

Theorem 1.1 (Cauchy)
For real vectors $a, b \in \mathbb{R}^n$, we have

$$\sum_{i=1}^n a_i b_i \leq \sqrt{\sum_{i=1}^n a_i^2} \sqrt{\sum_{i=1}^n b_i^2}$$

with equality if and only if a and b are linearly dependent.

Bunyakovsky [Bun59] showed a corresponding form using integrals. Later, Schwarz [Sch88] showed the following generalization for arbitrary inner products.

Theorem 1.2 (Schwarz)
For all vectors x and y in an inner product space, we have

$$|\langle x, y \rangle| \leq \|x\| \cdot \|y\| = \sqrt{\langle x, x \rangle}\sqrt{\langle y, y \rangle}$$

with equality if and only if x and y are linearly dependent.

1.6.2 The Inequalities of Rogers and Hölder

Rogers [Rog88] proved the following inequality.

Theorem 1.3 (Rogers)
For $0 < s < 1$ and nonnegative vectors $a, b \in \mathbb{R}^n$ ($a_i, b_i \geq 0$), we have

$$\sum_{i=1}^{n} a_i^s b_i^{1-s} \leq \left(\sum_{i=1}^{n} a_i \right)^s \left(\sum_{i=1}^{n} b_i \right)^{1-s} .$$

This is equivalent to the following inequality of Hölder [Höl89].

Theorem 1.4 (Hölder)
If p and q satisfy $(p-1)(q-1) = 1$ then we have for nonnegative numbers a_i, b_i

$$\sum_{i=1}^{n} a_i b_i \leq \left(\sum_{i=1}^{n} a_i^p \right)^{1/p} \left(\sum_{i=1}^{n} b_i^q \right)^{1/q} \qquad (p > 1) .$$

Equality occurs if and only if (a_i^p) and (b_i^q) are proportional, i.e., there are two numbers c_a and c_b, not both zero, such that $c_a a_i^p = c_b b_i^q$ for all i.
For $p < 1$, the inequality is reversed and equality occurs if either (a_i^p) and (b_i^q) are proportional or $(a_i b_i) = 0$.

Besides Cauchy's inequality, Hölder's inequality has several other interesting special cases, see [Bul03, p. 181]. The following is sometimes called Lyapunov's inequality [Lya01].

Theorem 1.5 (Lyapunov)
For all nonnegative vectors $a, b, p \in \mathbb{R}_{\geq 0}^n$ and $r, s, t \in \mathbb{R}$ with $0 < r < s < t$, we have

$$\left(\sum_{i=1}^{n} p_i a_i^s \right)^{t-r} \leq \left(\sum_{i=1}^{n} p_i a_i^r \right)^{t-s} \left(\sum_{i=1}^{n} p_i a_i^t \right)^{s-r} .$$

1.6.3 Chebyshev's Sum Inequality

Two n-tuples $(a_1, \ldots, a_n)$ and $(b_1, \ldots, b_n)$ of real numbers are called *similarly ordered* if $(a_i - a_k)(b_i - b_k) \geq 0$ for all $i, k \in [n]$. They are called *conversely ordered*[6] if $(a_i - a_k)(b_i - b_k) \leq 0$ for all $i, k = 1, \ldots, n$. The term similarly ordered is equivalent to the requirement that there is a permutation that transforms both tuples into nonincreasing sequences. In the same line, two tuples are conversely ordered if and only if there is a permutation that transforms one of the tuples into a nonincreasing

---

[6]also *oppositely ordered*, see [HLP59].

and the other tuple into a nondecreasing sequence. Below, we will use the same notation for n-dimensional real vectors $a, b \in \mathbb{R}^n$.

The following inequality is a consequence of an inequality for integrals that was published by Chebyshev [Che83].[7] Both are called Chebyshev's inequality.

Theorem 1.6 (Chebyshev)
If the vectors $a \in \mathbb{R}^n$ and $b \in \mathbb{R}^n$ are similarly ordered, then

$$\left(\sum_{i=1}^n a_i\right)\left(\sum_{i=1}^n b_i\right) \le n\sum_{i=1}^n a_i b_i \ .$$

The inequality is reversed if a and b are conversely ordered.

For $n > 0$, this is the same as the following relation between arithmetic means:

$$\frac{\sum_{i=1}^n a_i}{n} \cdot \frac{\sum_{i=1}^n b_i}{n} \le \frac{\sum_{i=1}^n a_i b_i}{n} \ .$$

More general versions are discussed in the survey of Mitrinović and Vasić [MV74], for instance the following variant that also appeared in a paper of Hayashi [Hay20].

Theorem 1.7 (Chebyshev)
For similarly ordered vectors $a \in \mathbb{R}^n$ and $b \in \mathbb{R}^n$ and any nonnegative vector $p \in \mathbb{R}_{\ge 0}^n$, we have

$$\left(\sum_{i=1}^n p_i a_i\right)\left(\sum_{i=1}^n p_i b_i\right) \le \left(\sum_{i=1}^n p_i\right)\left(\sum_{i=1}^n p_i a_i b_i\right) \ .$$

The inequality is reversed if a and b are conversely ordered.

If $p \in \mathbb{R}_{\ge 0}^n$ is nonzero, this corresponds to the following weighted arithmetic means relation:

$$\frac{\sum_{i=1}^n p_i a_i}{\sum_{i=1}^n p_i} \cdot \frac{\sum_{i=1}^n p_i b_i}{\sum_{i=1}^n p_i} \le \frac{\sum_{i=1}^n p_i a_i b_i}{\sum_{i=1}^n p_i} \ .$$

A direct consequence is the following. Given $a, b \in \mathbb{R}^n$ and $r \in \mathbb{R}$, suppose that a_i^r and b_i^r are defined within $\mathbb{R}$ for all $i \in [n]$ and that the corresponding tuples $(a_1^r, \dots, a_n^r)$ and $(b_1^r, \dots, b_n^r)$ are similarly ordered. Then we have

$$\frac{\sum_{i=1}^n p_i a_i^r}{\sum_{i=1}^n p_i} \cdot \frac{\sum_{i=1}^n p_i b_i^r}{\sum_{i=1}^n p_i} \le \frac{\sum_{i=1}^n p_i (a_i b_i)^r}{\sum_{i=1}^n p_i} \ .$$

One particular case where such inequalities can be obtained occurs for arbitrary r and nonnegative vectors a and b that are similarly or conversely ordered. Another special case is for *odd* integers r (or their reciprocals) and arbitrary real vectors a and b that are similarly or conversely ordered.

---

[7]Russian name Пафнутий Львович Чебышёв, transcribed in many different variants, e.g., as Čebyšev or Tchebychef[f]. A better English transcription would be Chebyshov.

1.6.4 Inequalities for Arithmetic, Geometric, Harmonic, and Power Means

For $1 \leq n \in \mathbb{N}$, consider an n-dimensional nonnegative vector $a \in \mathbb{R}_{\geq 0}^n$. Then several notions of means or averages can be considered. The most important ones are the *arithmetic mean*

$$\mathfrak{A}(a) = \underset{i \in [n]}{\mathfrak{A}} \, a_i := \frac{1}{n} \sum_{i=1}^{n} a_i \ ,$$

the *geometric mean*

$$\mathfrak{G}(a) = \underset{i \in [n]}{\mathfrak{G}} \, a_i := \sqrt[n]{\prod_{i=1}^{n} a_i} \ ,$$

and the *harmonic mean*

$$\mathfrak{H}(a) = \underset{i \in [n]}{\mathfrak{H}} \, a_i := \frac{n}{\sum_{i=1}^{n} \frac{1}{a_i}} \ .$$

All those means can be considered as special cases of the following power means. For $0 \neq r \in \mathbb{R}$, let the power mean of order r be defined as

$$\mathfrak{M}_r(a) := \left(\frac{1}{n} \sum_{i=1}^{n} a_i^r \right)^{1/r} \ .$$

(In the undefined case where $r < 0$ and there is some $a_i = 0$, the power mean is defined to be zero.) The arithmetic and the harmonic mean obviously correspond to $r = 1$ and $r = -1$, respectively. The case $r = 2$ is related to the length of the vector a by $\mathfrak{M}_2(a) = \frac{1}{\sqrt{n}} \|a\|$. For $r = 0$, the power mean is defined as the geometric mean: $\mathfrak{M}_0(a) := \mathfrak{G}(a)$. This is consistent with the observation that the geometric mean is the limit of the power means for $r \to 0$.

One of the most famous inequalities regarding mean values states that the geometric mean is bounded from above by the arithmetic mean. This is usually called the *inequality of arithmetic and geometric means*, or short the *AM-GM inequality*. Additionally, it is well known that the geometric mean is bounded from below by the harmonic mean. These inequalities can be generalized by the following statement for power means.

Theorem 1.8
For $q < r$ ($q, r \in \mathbb{R}$), the power means of any tuple or vector a with $a_i \geq 0$ satisfy

$$\mathfrak{M}_q(a) \leq \mathfrak{M}_r(a) \ .$$

Using a vector p of positive weights p_i, a weighted power mean can be defined as

$$\mathfrak{M}_{r,p}(a) := \left(\frac{\sum_{i=1}^{n} p_i \cdot a_i^r}{\sum_{k=1}^{n} p_k} \right)^{1/r} \ .$$

For the special case $r = 0$, the weighted power mean is defined as the weighted geometric mean:

$$\mathfrak{G}_p(a) := \mathfrak{M}_{0,p}(a) := \prod_{i=1}^{n} a_i^{p_i / \sum_{k=1}^{n} p_k} \ .$$

For $r = 1$ and $r = -1$, the weighted power means correspond to the usual definition of the weighted arithmetic mean $(\mathfrak{A}_p(a) := \mathfrak{M}_{1,p}(a))$ and the weighted harmonic mean $(\mathfrak{H}_p(a) := \mathfrak{M}_{-1,p}(a))$.

Theorem 1.9
For $q < r$ $(q, r \in \mathbb{R})$, the weighted power means for positive weights p_i and any tuple or vector a with $a_i \geq 0$ satisfy
$$\mathfrak{M}_{q,p}(a) \leq \mathfrak{M}_{r,p}(a) \ .$$

We will also use the following lemma.

Lemma 1.4
Given n arbitrary nonnegative numbers a_i and $p \in \mathbb{R}$, the following holds:

$$\left(\sum_{i=1}^{n} a_i \right)^p \geq n^{p-1} \cdot \sum_{i=1}^{n} a_i^p \qquad \textit{for } 0 \leq p \leq 1 \ ,$$

$$\left(\sum_{i=1}^{n} a_i \right)^p \leq n^{p-1} \cdot \sum_{i=1}^{n} a_i^p \qquad \textit{for } p \leq 0 \textit{ or } p \geq 1 \ .$$

Proof   The lemma corresponds to the inequalities

$$\left(\frac{\sum_{i=1}^{n} a_i}{n} \right)^p \geq \frac{1}{n} \sum_{i=1}^{n} a_i^p \qquad \text{and} \qquad \left(\frac{\sum_{i=1}^{n} a_i}{n} \right)^p \leq \frac{1}{n} \sum_{i=1}^{n} a_i^p \ .$$

The basic form of Jensen's inequality states that $f(\frac{1}{n} \sum_{i \in [n]} a_i) \leq \frac{1}{n} \sum_{i \in [n]} f(a_i)$ for any real convex function f and all $a \in \mathbb{R}^n$. The inequality is reversed if f is concave. The correctness of the lemma for $a_i \geq 0$ follows since the function $f(x) = x^p$ is concave for $0 \leq p \leq 1$ and convex for $p \leq 0$ or $p \geq 1$.

The inequalities of the lemma could also be written as $\mathfrak{A}(a)^p \geq \mathfrak{A}(a^p)$ and $\mathfrak{A}(a)^p \leq \mathfrak{A}(a^p)$. This can be generalized to the weighted arithmetic mean in the following way.

Lemma 1.5
Given n arbitrary nonnegative numbers a_i and $p \in \mathbb{R}$, the following holds:

$$\left(\sum_{i=1}^{n} x_i a_i \right)^p \geq \left(\sum_{i=1}^{n} x_i \right)^{p-1} \cdot \sum_{i=1}^{n} x_i a_i^p \qquad \textit{for } 0 \leq p \leq 1 \ ,$$

$$\left(\sum_{i=1}^{n} x_i a_i \right)^p \leq \left(\sum_{i=1}^{n} x_i \right)^{p-1} \cdot \sum_{i=1}^{n} x_i a_i^p \qquad \textit{for } p \leq 0 \textit{ or } p \geq 1 \ .$$

Proof   The lemma corresponds to the inequalities

$$\left(\frac{\sum_{i=1}^{n} x_i a_i}{\sum_{i=1}^{n} x_i}\right)^p \geq \frac{\sum_{i=1}^{n} x_i a_i^p}{\sum_{i=1}^{n} x_i} \qquad \text{and} \qquad \left(\frac{\sum_{i=1}^{n} x_i a_i}{\sum_{i=1}^{n} x_i}\right)^p \leq \frac{\sum_{i=1}^{n} x_i a_i^p}{\sum_{i=1}^{n} x_i} \ .$$

The weighted form of Jensen's inequality states that, for nonnegative weights x_i with $\sum_{i=1}^{n} x_i = 1$, the inequality $f(\sum_{i \in [n]} x_i a_i) \leq \sum_{i \in [n]} x_i f(a_i)$ holds true for any real convex function f and all $a \in \mathbb{R}^n$. The inequality is reversed if f is concave. The correctness of the lemma for $a_i \geq 0$ follows since the function $f(x) = x^p$ is concave for $0 \leq p \leq 1$ and convex for $p \leq 0$ or $p \geq 1$.

In the context of this book, it might also be interesting to use the following consequence that results from applying $x_i := y_i b_i^r$ and $a_i := b_i^s$.

$$\left(\frac{\sum_{i=1}^{n} y_i b_i^{r+s}}{\sum_{i=1}^{n} y_i b_i^r}\right)^p \geq \frac{\sum_{i=1}^{n} y_i b_i^{r+sp}}{\sum_{i=1}^{n} y_i b_i^r} \qquad \text{and} \qquad \left(\frac{\sum_{i=1}^{n} y_i b_i^{r+s}}{\sum_{i=1}^{n} y_i b_i^r}\right)^p \leq \frac{\sum_{i=1}^{n} y_i b_i^{r+sp}}{\sum_{i=1}^{n} y_i b_i^r} \ .$$

Lemma 1.6
Given n arbitrary nonnegative numbers a_i and $p \in \mathbb{R}$, the following holds:

$$\left(\sum_{i=1}^{n} x_i a_i^{r+s}\right)^p \geq \left(\sum_{i=1}^{n} x_i a_i^r\right)^{p-1} \cdot \sum_{i=1}^{n} x_i a_i^{r+sp} \qquad \text{for } 0 \leq p \leq 1 \ ,$$

$$\left(\sum_{i=1}^{n} x_i a_i^{r+s}\right)^p \leq \left(\sum_{i=1}^{n} x_i a_i^r\right)^{p-1} \cdot \sum_{i=1}^{n} x_i a_i^{r+sp} \qquad \text{for } p \leq 0 \text{ or } p \geq 1 \ .$$

Further information on elementary inequalities can be found in the well-known books by Hardy, Littlewood and Pólya [HLP59], Mitrinović [Mit64], Marcus and Minc [MM64], Mitrinović [Mit70], Beckenbach and Bellman [BB83], Mitrinović, Pečarić and Fink [MPF93], Bullen [Bul03] and Steele [Ste04].[8]

---

[8]Be aware that in some of the books, important requirements for the tuples or vectors (such as strict positivity) are defined only in the introductory part.

Chapter 2

Motivation

Die Zahlentheorie ist nützlich, weil man mit ihr promovieren kann.
(Number theory is useful, since one can graduate with it.)

Edmund Landau

In this work, we investigate entry sums of matrix powers or, more generally, quadratic forms as defined in the introduction. We present inequalities relating different entry sums of powers of a matrix to each other. An important special case occurs for adjacency matrices of graphs, where we obtain inequalities for different numbers of walks with certain lengths and specified sets of start and end vertices. In general, these inequalities are interesting for the analysis of various kinds of *iterative processes*. As can be seen from the description below, this includes not only combinatorial counting problems and extremal graph theory. There are close connections to quite different disciplines outside of discrete mathematics, for example to random walks and Markov chains, to the theory of formal languages and their representation by automata (combinatorics on words) in theoretical computer science, to population genetics in theoretical biology, to statistical mechanics, and to quantum field theory as well as quantum chemistry.

2.1   Simple Combinatorial Problems

To demonstrate the wide range of connections of k-step walks in graphs, we start with a toy example.

Move Sequences on a Chessboard.

How many different sequences consisting of exactly k moves could a king perform on a chessboard? Is it possible to find a formula for the number of those sequences? We assume that the board contains no other pieces, so castling moves can be ignored and the possible choices for the next move depend only on the current position of the king. The problem can be solved by constructing a graph where each square of the board is represented by a vertex and two vertices are connected by an edge if and only if a king is allowed to move between the corresponding squares. Since this relation is symmetric, the corresponding graph is undirected. Now the number of different sequences of k moves can be deduced from the entries of the k-th power of the adjacency matrix of the derived graph. For the total number of k-move sequences, this has been used by Cvetković [Cve70b] to deduce a formula for this number of sequences on a generalized $n \times n$ chessboard. In general, this adjacency matrix approach could also be used to count the move sequences starting and ending at certain squares, ranks, or files. Of course, it could also be applied to other chess pieces (with obvious restrictions in the case of a pawn, where the graph would be directed). As we will see, it is indeed possible to deduce a quite simple formula for the corresponding number of move sequences (using the spectral decomposition, see next chapter). Several generalizations are discussed in the book of Cvetković, Doob and Sachs [CDS79, p. 214].

Variations with Repetitions and Restrictions.

A more universal kind of application was described in another paper of Cvetković [Cve70c]. He considered the number of variations with repetitions (k-tuples of a set containing n elements), with the additional restriction that only certain pairs are allowed as consecutive elements. In a natural way, this corresponds to a restricted set of words of a fixed length k over a given alphabet of size n, a topic that is discussed below.

2.2 Automata and Formal Languages

Words Generated or Accepted by Automata.

Consider a formal language that can be described by some kind of *finite-state machine*, that is, an automaton that may read an input string and, where applicable, write an output string. Assume that the language under consideration is either the set of accepted input words or the set of generated output strings. To this end, an initial (or start) state and a set of accepting (or final) states might be specified. We assume that the automaton reads the input (or writes the output) symbol-by-symbol, one character in each step, where every step corresponds to a state transition and the input symbol (together with the current state) determines the next state. If the automaton creates an output string, the output character could depend solely on the current state (i.e., the output symbols are state labels like in a *Moore machine*), or it could depend additionally on the input character (i.e., the output symbols are transition labels like in a *Mealy machine*). Under the assumption that all output labels are different, there is a natural bijection between the set of outputs of the automaton and the set of walks through the automaton graph. In the case of the Moore automaton, the walk is described by the sequence of state labels (associated with the vertices of the automaton graph), while for the case of the Mealy automaton, the walk is described by the sequence of transition labels (associated with the edges of the automaton graph). In the case of a Mealy automaton, it would be sufficient to have different output labels for the edges leaving each state to distinguish the different walks through the graph. A similar correspondence applies if we consider a formal language that is accepted by a *deterministic finite automaton*. A conceivable extension of the models above is the introduction of several possible initial states. Then, in the cases of Mealy automata and deterministic finite automata, a sufficient criterion for distinguishing the possible walks through the state graph would be to have disjoint edge label sets for the start states.

If we assume some knowledge about the underlying graph of the automaton (and maybe about the set of initial states and the set of accepting states), we would like to deduce useful bounds on the number of different words that are accepted (or generated) by this automaton. To this end, we would apply the introduced setting of bilinear forms using adjacency matrix powers and characteristic vectors for the sets of start states and final states.

Bounds on the number of words of length exactly k (or at most k) in a certain language can be interesting if a certain kind of density of formal languages is considered. Such a concept is mentioned in the book of Salomaa and Soittola [SS78, p. 92]. They note that density results can be used to prove that a given language does not belong to a given family of languages and they provide an example.

For an introduction and general treatment of combinatorics on words, see Lothaire [Lot83], Perrin [Per01], Berstel and Perrin [BP07], and Flajolet and Sedgewick [FS09, Ch. V.5]. Relations between walks and words were also considered in the article of Bruns, Herzog and Vetter [BHV94].

Computation Paths.

Another application of our results is found in symmetric models of computation, which exhibit undirected configuration graphs. One particular example for such a model is the symmetric Turing machine, which was defined by Lewis and Papadimitriou [LP82] to introduce the complexity class Symmetric Logspace (SL). They characterized SL by showing that all these problems are log-space reducible to undirected s,t-connectivity (USTCON). In this context, the number of computation paths consisting of k transitions equals the number of walks of length k in the corresponding configuration graph starting at the initial configuration. Assuming that the configuration graph is finite, it is also interesting to investigate the total number of different computation path segments of certain lengths starting at arbitrary vertices. Bounds could be given in terms of the number of configurations, total number of transitions, number of transitions incident to each configuration, and so on. Other bounds could take into account the number of computation path segments of other lengths.

2.3 Graph Density, Maximum Clique and Densest k-Subgraph

A particularly useful concept is the statistical notion of (sub)graph density. The usual definition applies to undirected graphs without loops, but similar definitions can be made for directed graphs and graphs that may contain loops. In any case, the density is defined as the ratio between the number of present edges and the maximum possible number of edges in any graph on the same number of vertices. In a similar way, the number of k-step walks induces a density of order k (by dividing the number of k-step walks in the graph by the maximum possible number of k-step walks in any graph on the same number of vertices, see Kosub [Kos05]). In this book, we will use several inequalities for the number of k-step walks to deduce similar inequalities for k-th order densities. As special cases, we obtain statements about the classical notion of (sub)graph density.

Density concepts frequently occur in the context of Szemerédi's Regularity Lemma (see, e.g., the survey by Komlós and Simonovits [KS96]) and in extremal graph theory which is discussed in Section 2.4.

An initial motivation to investigate the number of k-step walks in graphs was the question which density functions allow polynomial time algorithms for detecting subgraphs of this density having a certain size and which density functions lead to NP-hard decision problems (see Holzapfel, Kosub, Maaß and Täubig [HKMT06]). In particular, our attention was attracted by the article of Feige, Kortsarz and Peleg [FKP01] (an improved version of an earlier paper by Kortsarz and Peleg [KP93]) on approximating the Densest k-Subgraph Problem (a problem introduced by Feige and Seltser [FS97]) and by the article of Alon, Feige, Wigderson and Zuckerman [AFWZ95] on derandomized graph products. In these papers, a couple of observations regarding the number of walks were used. The aim of the paper by Feige et al. was to find a dense subgraph in a graph that was given as an input together with the size of the subgraph. Since this problem is NP-hard (e.g., by reduction from Clique), an approximation algorithm was developed. They used the following lemma.

Lemma 2.1 (Feige, Kortsarz and Peleg)
In every undirected graph $G = (V, E)$ on n vertices having average degree $\overline{d}$, there exist two vertices $v_i, v_j \in V$ such that

$$\overline{d}^k / n \leq w_k(v_i, v_j) \ .$$

In the proof, Feige et al. remark that this lemma also follows from the more global fact that the total number of walks of length ℓ in a graph of average degree $\overline{d}$ is at least $n\overline{d}^\ell$. For a partial proof, they referred to the paper by Alon et al. (that only covers the case where k is *even*) and to a private communication with Noga Alon. However, as we will discuss later, it had been noticed already several years before in a paper by Erdős and Simonovits [ES82] that this global inequality can be proven for even and odd k using the results of Mulholland and Smith [MS59], Blakley and Roy [BR65], and London [Lon66a]. Also the result of Scheuer and Mandel [SM59] implies the global statement.

The paper of Alon, Feige, Wigderson and Zuckerman [AFWZ95] considered so called *derandomized graph products*, a deterministic version of *randomized graph products*. These had been introduced by Berman and Schnitger [BS92] and Blum [Blu91] as a randomized reduction from approximating MAX-SNP problems within constant factors arbitrarily close to 1 to approximating the size of a largest clique within a factor of n^ϵ. Alon et al. derandomized the reduction by using random walks on expander graphs. The main result of the paper is a lower bound for the probability that all steps of a random walk stay within a certain vertex set.

Another notion of density was considered by Andersen and Cioabă [AC07]. They investigated approximate solutions to the search for subgraphs of limited size with large spectral radius.

2.4   The Number of Paths and Extremal Graph Theory

A popular subject in graph theory is the search for the maximum number of edges $\mathrm{ex}(n, \mathbf{L})$ for any graph on n vertices such that it does not contain a graph from a given (finite or infinite) class $\mathbf{L}$ of forbidden subgraphs. These numbers are also called Turán numbers. The closely related Zarankiewicz number is the maximum number of edges in an $\mathbf{L}$-free *bipartite* graph on n vertices. Erdős and Simonovits [ES82] investigated so-called *compactness* results asserting for a given family $\mathbf{L}$ the existence of a much smaller $\mathbf{L}^* \subseteq \mathbf{L}$ with $\mathrm{ex}(n, \mathbf{L}^*) \approx \mathrm{ex}(n, \mathbf{L})$. In particular, they investigated the case where $\mathbf{L}$ consists of cycles.

Using the Erdős-Stone theorem [ES46], it was shown by Erdős and Simonovits [ES66] that, surprisingly, the asymptotic growth of $\mathrm{ex}(n, \mathbf{L})$ is dictated in some sense by the minimum of the chromatic numbers $\chi(L)$ of the graphs $L \in \mathbf{L}$:

$$\mathrm{ex}(n, \mathbf{L}) = \left(1 - \frac{1}{\min_{L \in \mathbf{L}} \chi(L) - 1} + o(1)\right)\binom{n}{2}.$$

This distinguishes the case where $\mathbf{L}$ contains a bipartite graph from the other cases. Since a bipartite graph L has $\chi(L) = 2$, this implies $\mathrm{ex}(n, \mathbf{L}) = o(n^2)$ and this case is called *degenerate*. For a forbidden graph L^* of minimum chromatic number, the asymptotic formula implies $\mathrm{ex}(n, L^*) = \mathrm{ex}(n, \mathbf{L}) + o(n^2)$. While this implies $\frac{\mathrm{ex}(n, L^*)}{\mathrm{ex}(n, \mathbf{L})} \to 1$ for $n \to \infty$ in the nondegenerate case, this is not necessarily true for degenerate problems. Therefore, Erdős and Simonovits [ES82] were interested in compactness theorems for degenerate problems that assert for some $\mathbf{L}$ the existence of a much smaller $\mathbf{L}^* \subseteq \mathbf{L}$ with $\frac{\mathrm{ex}(n, \mathbf{L}^*)}{\mathrm{ex}(n, \mathbf{L})} \to 1$ for $n \to \infty$.

Erdős and Simonovits further explain that there are many extremal graph problems where it is shown that a graph G on n vertices contains some forbidden $L \in \mathbf{L}$ by choosing another family $\mathbf{P}$ and proving that G contains many $p \in \mathbf{P}$. Assume that $G = (V, E)$ has $|E| = \mathrm{ex}(n, \mathbf{P}) + k$ for $k > 0$. Such a graph G is called *supersaturated* and clearly must contain a forbidden subgraph. Somewhat surprisingly, it must contain *many* of those forbidden subgraphs. In this context, Erdős and Simonovits were interested in lower bounds for the number of paths P_k on $k + 1$ vertices and

k edges[1] in a graph. Now the problem is that the number of paths of a certain length in a graph is not that easy to count. It is much easier to count the number of walks since one does not have to care about the repetitions of edges and vertices. To this end, it was shown in [ES82] that "in some sense the difference of the number of walks and paths is negligible".

Lemma 2.2 (Erdős and Simonovits)
Let $q < 1 + \frac{1}{k-1}$ and assume that the maximum degree is bounded by $\Delta \leq \bar{d}^q$. If w_k^* denotes the number of k-walks which are not (directed) paths, then

$$w_k^* = o(w_k) \quad for \quad \bar{d} \to \infty \ .$$

This was used to prove the following theorem.

Theorem 2.1 (Erdős and Simonovits)
Suppose that $f(n, \bar{d})$ is the minimum number of k-walks in a graph on n vertices having average degree $\bar{d}$. Then all graphs on n vertices with average degree $\bar{d}$ contain at least $\frac{1}{2}f(n, \bar{d}) - o(f(n, \bar{d}))$ paths consisting of $k + 1$ vertices and k edges for $\bar{d} \to \infty$.

The 1/2-factor compensates for the fact that each path yields two walks in opposing directions.

Walks, paths, and graph homomorphisms for *trees* were investigated in the articles of Csikvári [Csi10], Bollobás and Tyomkyn [BT12], and Csikvári and Lin [CL14].

Bounds on the number of walks and the number of paths are also used in the article by Keevash, Sudakov and Verstraëte [KSV13]. Also closed walks are considered.

Lemma 2.3 (Keevash et al.)
Suppose G is a bipartite graph on n vertices with part sizes $n/2 + o(n)$, maximum degree Δ and girth at least $2\ell + 2$. Then

$$\frac{cl_{2\ell+2}}{n} < (1/2 + o(1))n\Delta^2 + (4\Delta)^{\ell+1} \ .$$

Keevash et al. show a theorem for nonreturning walks that is very similar to the theorem for general walks that Erdős and Simonovits deduced from the theorem of Mulholland and Smith. This is used to show connections between Turán numbers for certain sets of forbidden cycles to the related Zarankiewicz numbers.

Some of the recent developments in extremal graph theory (in particular, in spectral extremal graph theory) can be found in the survey of Nikiforov [Nik11].

---

[1]Note that Erdős and Simonovits used a slightly different notation in their paper. Therein, a path P_k and a walk W_k are defined to have k vertices instead of k edges. Also, the number of walks w_k was implicitly divided by n.

2.5   Means, Variances, and Irregularity

Collatz and Sinogowitz [CS57] noticed that the average degree of a graph is always less than or equal to the graph index ($\overline{d} \leq \lambda_1$). Equality is attained only if the graph is regular. Therefore, they proposed to use $\lambda_1 - \overline{d}$ as a measure for the deviation of the degrees from regularity.

This measure has been compared by Bell [Bel92] to the *degree variance* of a graph G, which is defined as

$$\text{Var}(G) := \frac{1}{n} \sum_{v \in V} \left(d_v - \overline{d} \right)^2 \ .$$

Albertson [Alb97] defined the *irregularity* $irr(G)$ of a graph G as

$$irr(G) := \sum_{\{v,w\} in E} |d(v) - d(w)| \ .$$

However, note that $irr(G) = 0$ does not imply that G is regular. This implication holds true only for the connected components.

In recent publications, a lot of interest was attracted by questions related to these irregularity measures, see, e.g., Hollas [Hol05; Hol06], Nikiforov [Nik06a], Zhou and Luo [ZL08], de Oliveira, Oliveira, Justel and de Abreu [dOOJdA13], Fath-Tabar, Gutman and Nasiri [FGN13], Elphick and Wocjan [EW14], Goldberg [Gol14a; Gol14b; Gol15], Hamzeh and Réti [HR14] and Gutman, Furtula and Elphick [GFE14].

Since

$$0 \leq \sum_{v \in V} \left(d_v - \overline{d} \right)^2 = \sum_{v \in V} \left(d_v^2 - 2 d_v \overline{d} + \overline{d}^2 \right) = w_2 - 2m \frac{4m}{n} + n \frac{4m^2}{n^2} = w_2 - \frac{w_1^2}{n} \ ,$$

we have $0 \leq \text{Var}(G) = \frac{w_2}{n} - \left(\frac{w_1}{n} \right)^2$. In this form, the average number of 2-step walks per vertex is compared to the squared average number of 1-step walks. This yields a very simple proof for $w_1^2 \leq w_0 w_2$ or $\overline{d} w_1 \leq w_2$. In principle, this inequality can also be found (without reference to walks) in the slightly different form $1 + c_v^2 = \frac{n}{4m^2} \sum_{i=1}^{n} d_i^2$ within the paper of Edwards [Edw77], where $c_v \in \mathbb{R}_{\geq 0}$ could be called "the vertex degree coefficient of variation" for the graph. The closely related inequality $irr(G)^2 \leq m(nM_1 - 4m^2) = m(nw_2 - w_1^2)$ has been shown by Zhou and Luo [ZL08].

In general, the method above can be used to show $w_k^2 \leq w_0 w_{2k}$, which is a special case of inequalities that we will discuss later (see Theorems 4.1 and 4.2):

$$0 \leq \sum_{v \in V} \left[w_k(v) - \frac{w_k}{n} \right]^2 = \sum_{v \in V} \left[w_k(v)^2 - 2 w_k(v) \frac{w_k}{n} + \left(\frac{w_k}{n} \right)^2 \right]$$

$$= w_{2k} - 2 \frac{w_k^2}{n} + n \frac{w_k^2}{n^2} = w_{2k} - \frac{w_k^2}{n}$$

$$w_k^2 \leq w_0 w_{2k} \ .$$

Under certain conditions, similar results can be obtained for directed graphs, see Section 6.1.

2.6 Random Walks and Markov Chains

There is a strong connection between walks on graph structures and probability theory in the context of Markov chains. While the classical results of random walks focused on walking on simple and well-structured infinite graphs like grids, the more recent research focused on finite graphs having an arbitrary, or at least less restricted, structure.

A *random walk* on a (directed or undirected) graph is a model of a stochastic process that is defined as follows. Assuming that the random walk commences at a certain starting vertex, an outgoing edge of the current vertex is selected at random and the current position is moved to the vertex at the opposing end of the selected edge. This model can be defined for directed or undirected graphs, as well as for weighted or unweighted graphs in an obvious way. For unweighted graphs, the next vertex is chosen *uniformly* at random from the out-neighbors whereas for graphs with nonnegative edge weights, the probabilities of moving to a neighbor is proportional to the respective edge weight.

While the outcome of one particular run of this stochastic process is a single walk in the graph, the more interesting questions arise from a setting where a certain initial probability distribution $v_0 \in \mathbb{R}^n_{\geq 0}$ with respect to the starting vertex is assumed. Applying the different possible moves from a vertex v_i to a vertex v_j with their respective probabilities p_{ij}, the new probability distribution of the current vertex is $v_1^T = v_0^T P$, where P denotes the matrix that consists of the probabilities $p_{ij} = w_{ij} / \sum_{k=1}^n w_{ik}$. Here, w_{ij} is the weight of the edge (v_i, v_j), or zero if there is no such edge. For unweighted graphs, we have $p_{ij} = 1/d_{\text{out}}(v_i)$ if there is an edge (v_i, v_j). In the undirected case, this obviously simplifies to $p_{ij} = 1/d(v_i)$.

For each subsequent step $i + 1$, another probability distribution $v_{i+1} \in \mathbb{R}^n_{\geq 0}$ for the current vertex results from the former probability distribution v_i according to the law $v_{i+1}^T = v_i^T P$. Therefore, we have

$$v_k^T = v_0^T P^k \qquad \text{for all } k \in \mathbb{N}.$$

Now, the most interesting questions are: What is the expected number of steps for the first visit of a given vertex, for the first return to the starting point, or until we have visited all vertices? How fast does the probability distribution converge to its limit?

This model of random walks is equivalent to the model of finite Markov chains. The general case corresponds to directed graphs with nonnegative edge weights. While time-reversible Markov chains correspond to random walks on undirected graphs, the case of symmetric Markov chains corresponds to random walks on regular undirected graphs.

Detailed discussions of random walks on graphs can be found in the articles of Göbel and Jagers [GJ74], Lawler [Law86], Broder and Karlin [BK89], Palacios [Pal90], Feige [Fei95b; Fei95a] and Elsässer and Sauerwald [ES11]. Random walks were also considered on restricted graph classes like trees as, for instance, in the paper by Pearce [Pea80]. An overview of the subject can be found in the survey of Lovász [Lov96] and in the (unfinished) book by Aldous and Fill [AF02].

Also, most of the books on nonnegative matrices, contain more or less detailed collections of results for finite Markov chains; see for instance Seneta [Sen73, Chapter 4], Graham [Gra87, Chapter 6], Minc [Min88, Chapter 6], Berman and Plemmons [BP94, Chapter 8].

2.7 Largest Eigenvalue

As we will see, there is a close connection between the growth of the number of k-step walks for increasing k and the largest eigenvalue of the adjacency matrix (or, more precisely, the eigenvalues of largest absolute value). This opens up an opportunity for a quite universal approach to walk-related bounds that exploits the various relationships of the largest eigenvalue λ_1 of adjacency matrices to the diverse properties of graphs. To this end, we will derive lower bounds for λ_1 in terms of the number of walks. In turn, λ_1 can be used to bound other important graph measures.

For the chromatic number χ of a graph, Hoffman [Hof70] obtained the bound

$$1 - \lambda_1/\lambda_n \leq \chi \ ,$$

relating χ to the ratio of the largest and the smallest eigenvalue.

The clique number ω of a graph can be bounded using an inequality of Wilf [Wil86]:

$$n/(n - \lambda_1) \leq \omega \qquad \text{or} \qquad \lambda_1 \leq (1 - 1/\omega)n \ .$$

Another interesting application of λ_1 considers the SIS model of disease spreading, where a susceptible (S) individual is possibly infected by an already infected (I) neighbor, and subsequently may become cured again. If an infected individual infects a certain neighbor with probability β and is cured with probability δ then the expected size of the infected part of the population reduces exponentially if $\beta/\delta < 1/\lambda_1$, i.e., $1/\lambda_1$ is the epidemic threshold in this model (see Ganesh, Massoulié and Towsley [GMT05] and Chakrabarti et al. [CWW$^+$08]). Besides the spreading of viruses in biological and computer networks, this model can also be applied to rumor spreading and information broadcasting.

Furthermore, bounds for the spectral radius are particularly important for bounding the spectral gap, i.e., the distance to the second largest eigenvalue or the eigenvalue with second largest absolute value. This is an indispensable ingredient for analyzing expanders (see, e.g., Jukna [Juk11]) and the convergence rate of Markov chains.

More information on applications of graph spectra can be found in the papers by Mohar and Poljak [MP93], Cvetković and Gutman [CG09], Estrada [Est09], Cvetković and Gutman [CG11], Cioabă [Cio11], Van Mieghem [Van11] and Spielman [Spi12].

2.8 Population Genetics and Evolutionary Theory

Consider the following simplified theoretical model for a large population of diploid individuals where we have a single locus with n alleles $L_1, \ldots, L_n$. Let $a_{ij} = a_{ji} \geq 0$

denote the relative viability for individuals of the genotype L_iL_j. If p_i is the frequency of L_i in a certain generation (i.e., $\sum_{i=1}^{n} p_i = 1$) then the mean viability assuming random mating is $\sum_{i=1}^{n} \sum_{j=1}^{n} a_{ij} p_i p_j$. In the next generation, the new frequency p'_k of L_k is

$$p'_k = \frac{p_k \sum_{j=1}^{n} a_{kj} p_j}{\sum_{i=1}^{n} \sum_{j=1}^{n} a_{ij} p_i p_j}$$

and the mean viability of the next generation is $\sum_{i=1}^{n} \sum_{j=1}^{n} a_{ij} p'_i p'_j$. For biological reasons, Mandel and Hughes [MH58] made the conjecture that the mean viability cannot decrease from one generation to the next generation.[2] If we interpret the frequencies p_i as a vector p and if A denotes the symmetric matrix of the relative viabilities a_{ij}, this conjecture corresponds to the following inequality:

$$\mathrm{sum}_p(A) \le \frac{\mathrm{sum}_p(A^3)}{(\mathrm{sum}_p(A))^2} \qquad \text{or} \qquad (\mathrm{sum}_p(A))^3 \le \mathrm{sum}_p(A^3) \ .$$

The conjecture of Mandel and Hughes [MH58] was proven at nearly the same time by Scheuer and Mandel [SM59], Mulholland and Smith [MS59] as well as Atkinson, Watterson and Moran [AWM60]. Hence, in the model described above, a population that starts in any state where all possible alleles are present will tend monotonically to a state of maximum mean viability. A more direct proof for the result of Atkinson et al. was given by Kingman [Kin61b].

Further results and remarks, also on the misinterpretation of Fisher's fundamental theorem of natural selection in the context of discrete-generation models, can be found in the articles by Kingman [Kin61a], Edwards [Edw67], Mandel [Man68], Deakin [Dea73], Mandel [Man80], Edwards [Edw94], Ewens and Watterson [EW10] and Bürger [Bür11], as well as in the book by Edwards [Edw00].

2.9 Theoretical Chemistry

In the field of theoretical chemistry, several measures were derived from graphs representing the carbon skeleton of molecules. Some of those measures show clear correlations with physico-chemical properties of the corresponding molecules. Two approaches that used such measures were published as a reaction to articles of Taylor, Pignocco and Rossini [TPR45] and Willingham, Taylor, Pignocco and Rossini [WTPR45] where certain properties of hydrocarbons are derived from contributions of the different component parts of a molecule and from contributions of the interactions between adjacent component parts. Wiener [Wie47; Wie48] proposed to approximate the boiling points of alkane isomers by a formula that depends on two parameters: the sum of distances of all vertex pairs in the graph representing the molecule structure (the 'path number', today called the Wiener index) and the number of vertex pairs having distance equal to 3 in the graph (the 'polarity number'). Note that the graphs corresponding to the respective molecules (alkanes) are trees.

---

[2]Note that generations are discrete and non-overlapping in this model. For other cases where generations overlap or when selection is infinitesimally slow, it was shown before (by introducing a continuous time parameter) that mean viability cannot decrease.

For trees, there is only one path connecting a given pair of vertices. Thus, the polarity number corresponds to the number of undirected paths of length 3 in the graph. Platt [Pla47; Pla52] proposed another approach for the approximation that led to the definition of the Platt index to be the sum of first $C-C$ neighbors of every $C-C$ bond in the molecule, i.e., the sum over all edges of the corresponding numbers of incident edges in the representing graph. Note that this measure corresponds to the number of directed paths of length 2.

A similar measure was proposed later by Gordon and Scantlebury [GS64]. For this descriptor (sometimes called Gordon-Scantlebury index), it was proposed to count the different ways of possible embeddings of a path of length 2 into the carbon skeleton of the molecule. This is the number of undirected paths of length 2 in the graph and thus just half of the Platt index.

Another approach concerning the abstraction of molecules as graphs that focused on the characteristic polynomial of its adjacency matrix was proposed by Spialter [Spi64a; Spi64b].

For branched hydrocarbons, two very popular measures derived from the vertex degrees of the underlying graphs originated in the Seventies of the last century. Since they resulted from the work of a group of people working at an institute in Zagreb, they were later called the *Zagreb group indices*, or just *Zagreb indices*. The first explicit definition of those indices appeared in the paper by Gutman, Ruščić, Trinajstić and Wilcox [GRTW75]. Erroneously, it referred to the earlier article by Gutman and Trinajstić [GT72] as the point where these measures where introduced. Actually, this is not true. (This historical development is clarified in the recent paper by Gutman [Gut14b].) Nevertheless, many papers in this area refer to the earlier paper [GT72] as the first appearance of the Zagreb indices. The first and the second *Zagreb [group] index* for an undirected graph $G = (V, E)$ are defined as

$$M_1 = \sum_{v \in V} d_v^2 \qquad \text{and} \qquad M_2 = \sum_{\{x,y\} \in E} d_x d_y \ .$$

For a long time, it has not been recognized that obviously

$$M_1 = w_2 \qquad \text{and} \qquad M_2 = w_3/2 \ .$$

During all the decades of research on topological indices, it had been overlooked that two of the most popular descriptors were in fact just special cases of measures defined by the number of walks. Only recently, it was observed by Nikolić, Kovačević, Miličević and Trinajstić [NKMT03] and Braun, Kerber, Meringer and Rücker [BKMR05] that M_1 equals w_2 (which is implicitly contained in the paper by Gutman, Rücker and Rücker [GRR01], but not explicitly stated there) and that M_2 equals $w_3/2$. In this context, it must be criticized that many authors (and also whole scientific communities) do not really put enough effort into searching for related articles and results. Another example is the alternative equation

$$M_1 = \sum_{\{x,y\} \in E} d(x) + d(y)$$

for the first Zagreb index. In contrast to the current perception in parts of the theoretical chemistry community, this simple relation was known (within the graph theory community) several decades before (see, e.g., Lovász [Lov79, Solutions 10.30 and 10.33], or de Caen [dCae98] or Das [Das04]). Later, we will see that some of the results that were obtained for the Zagreb indices can be generalized using the number of walks approach.

As already mentioned, Wiener's polarity number in trees is the same as the number of undirected paths of length 3. In trees (or more generally in triangle-free graphs), this is equal to $\sum_{\{x,y\} \in E} [d(x) - 1][d(y) - 1]$, a term that is sometimes referred to as the "reduced second Zagreb index". In general graphs, the number represented by this formula can be split into the number of simple 3-edge paths and three times the number of triangles (3-edge cycles). Relations between these measures are discussed in the articles of Došlić et al. [DFG⁺11], Andova, Cohen and Škrekovski [ACŠ12], Milošević, Réti and Stevanović [MRS12], Gutman and Réti [GR14], Gutman, Furtula and Elphick [GFE14], and Furtula, Gutman and Ediz [FGE14]. Properties of Zagreb indices have been reviewed by Nikolić, Kovačević, Milićević and Trinajstić [NKMT03] as well as Gutman and Das [GD04]. Degree-based topological indices were also reviewed and compared by Gutman [Gut13].

In an attempt to compute unique identifiers of chemical structures for a computer-based chemical information system, an article of Morgan [Mor65] introduced the notion of extended connectivity in chemical graphs. At the same time, a paper of Penny [Pen65] also considered the related problem of efficiently searching chemical structures. An approach based on indirect neighborhood was proposed there. Concerning Morgan's extended connectivity concept, Razinger [Raz82] showed that it is actually the same as the number of walks in the corresponding graph. Later, Razinger [Raz86] compared walk-based chemical structure indices to other known numerical structure descriptors. He showed that walk-based indices are excellent in discriminating similar nonisomorphic structures.

The possibilities of exploiting correlations between structural indices and the physico-chemical properties (e.g., to calculate orderings of isomers) were investigated by Randić, Hansen and Jurs [RHJ88], Balaban [Bal92] and Vukičević and Gašperov [VG10]. Based on the work of Skorobogatov and Dobrynin [SD88] and Diudea, Minailiuc and Balaban [DMB91], the so-called *layer matrices* and walk-related indices were investigated in the papers of Diudea [Diu94], Diudea, Topan and Graovac [DTG94] and Diudea [Diu96].

Other indices based, for example, on the eigenvalues of the edge adjacency matrix or on line graphs were considered in the articles of Gutman and Estrada [GE96], Estrada and Ramírez [ER96], Estrada [Est96; Est97; Est98] and Estrada, Guevara and Gutman [EGG98].

While most of those previously mentioned papers dealt with walks that were restricted in a certain way (like closed walks, paths and others), the number of *all* walks was used as a measure in the papers of Rücker and Rücker [RR93; RR00; RR01], Gutman, Rücker and Rücker [GRR01] and Lukovits, Milićević, Nikolić and Trinajstić [LMNT02]. Some of these papers also investigated the potential of using not only the total number of walks, but the numbers of walks starting at the different single vertices (atoms).[3] Also, walks of negative order were considered by Lukovits and Trinajstić [LT03] and Rücker and Rücker [RR03].

The differences between the conceptually similar approaches of classical random walks and equally probable walks of a predefined length are analyzed in the paper by Klein, Palacios, Randić and Trinajstić [KPRT04].

Another remarkable application of (spectral) graph theory is the close relation between the molecular orbital energy levels of the π-electrons in conjugated hydrocarbons and the eigenvalues of the adjacency matrix of the underlying skeleton graphs (in particular, between the total π-electron energy and the *graph energy*). We will pursue this matter further in the last part of this book (see Section 8.3).

2.10   Iterated Line Digraphs

The first appearances of the concept of line graphs that we are aware of are the articles by Whitney [Whi31; Whi32] and Krausz [Kra43]. While these articles only consider undirected graphs, we will focus on the slightly different directed variant, the concept of which seems to appear for the first time in an article of de Bruijn [dBru46], where it is applied to certain special digraphs. The term *line graph* and the formal extension to the *line digraph* in the general case are due to Harary and Norman [HN60]. Many other names were used as well, for instance, *interchange graph* (see van Rooij and Wilf [vRW65]), *arc digraph* (see Knuth [Knu67]), *derived graph* (see Beineke [Bei68]), or *adjoint graph* (see Sabidussi [Sab68]). The early results were summarized in the survey by Grünbaum [Grü69].

The *directed line graph* or *line digraph* $\mathcal{L}(G)$ of a directed graph $G = (V, E)$ is the directed graph that has a vertex v_e for each edge e of G where two vertices v_{e_1} and v_{e_2} of the line digraph representing the original edges $e_1 = (s_1, t_1)$ and $e_2 = (s_2, t_2)$ are connected by an edge (v_{e_1}, v_{e_2}) in the line digraph if and only if $t_1 = s_2$. In this way, the vertices of the line digraph correspond to the walks of length 1 in the original graph G and the edges of $\mathcal{L}(G)$ represent the walks of length 2 in G. The line digraph of an undirected graph is defined as the line digraph of its corresponding bidirected graph (using antiparallel edges for each undirected edge). Note that the line digraph of an undirected graph G is closely related but *not equal* to the bidirected line graph of G.

For an edge-weighted graph, a *weighted line digraph* can be defined as follows. We simply assume that the vertices v_{e_1} and v_{e_2} of the line digraph representing the original edges $e_1 = (s_1, t_1)$ and $e_2 = (s_2, t_2)$ with weights $w(e_1)$ and $w(e_2)$ are connected by an edge (v_{e_1}, v_{e_2}) of weight $w(e_1) \cdot w(e_2)$ in the weighted line digraph.

The repeated application of the line digraph operation yields the *k-th iterated line digraph*. For convenience, we define $\mathcal{L}_0(G) := G$. Starting with $\mathcal{L}_1(G) := \mathcal{L}(G)$, the definition continues recursively using $\mathcal{L}_{k+1}(G) := \mathcal{L}(\mathcal{L}_k(G))$.

Line digraphs appear in the papers of Kasteleyn [Kas63], Chartrand and Stewart [CS66], Beineke [Bei68], Geller and Harary [GH68], Hemminger and Beineke [HB78],

---

[3]Their notations $\text{awc}_k(i)$ and mwc_k correspond to our notations $w_k(v_i)$ and w_k.

Fiol, Alegre, Yebra and Fàbrega [FAYF85], Fiol and Lladó [FL92], Kwapisz [Kwa96], Snellman [Sne08], Bidkhori and Kishore [BK11] and de Freitas, Bonifácio, Robbiano and San Martín [dFBRS14].

It is clear that the line digraph has a Hamiltonian cycle if and only if the original graph has an Eulerian cycle. $\mathcal{L}(G)$ is Hamiltonian if and only if G is Eulerian, see Aigner [Aig67] and Kasteleyn [Kas67]. Kasteleyn [Kas67] also showed that the number of directed Hamiltonian cycles in $\mathcal{L}(G)$ equals the number of directed Eulerian walks in G. (Both results follow also from the one-to-one correspondence between walks and vertices or edges in iterated line graphs.) This was used by Sysło [Sys73] to show that the Traveling Salesman Problem (TSP) is solvable for line digraphs under certain conditions. Valdes, Tarjan and Lawler [VTL79] and Sysło [Sys82] proposed efficient algorithms to recognize line digraphs and to compute the root graph of a line digraph.

The most prominent examples for graph classes that are constructed using line digraph iteration are the *de Bruijn graphs* and the *Kautz graphs*. The de Bruijn graphs became well-known[4] after the articles of de Bruijn [dBru46] and Good [Goo46]. They can be constructed using line digraph iteration starting with a graph on just one vertex which has two loops. The vertices in these graphs represent all binary sequences of a certain length. Two vertices u and v are connected by a directed edge (u, v) if and only if the binary representation of v can be obtained by shifting the binary sequence of u by one position to the left (ignoring the first bit) and appending an arbitrary bit at the end. Hence, these graphs represent overlaps in binary sequences. The same principle can be applied if an alphabet consisting of more than two symbols is used. By taking a Hamilton path in the d-dimensional (or, equivalently, an Euler path in the $(d - 1)$-dimensional) de Bruijn graph, it is possible to construct so-called *de Bruijn sequences*. These are cyclic sequences where every word of a fixed length over a given alphabet occurs exactly once. The de Bruijn graphs for alphabet size a can also be seen as a combinatorial model of the Bernoulli map $x \mapsto ax \bmod 1$ in the context of dynamical systems theory. For instance, binary de Bruijn graphs can be used to model the Lorenz attractor. Further applications of de Bruijn graphs are routing (see Esfahanian and Hakimi [EH85], Samatham and Pradhan [SP89; SP91], Bermond and Peyrat [BP89], Sivarajan and Ramaswami [SR94] and Bermond and Fraigniaud [BF94]), broadcasting (see Heydemann, Opatrny and Sotteau [HOS92]) the design of dynamic networks and the construction of distributed hash tables (see Naor and Wieder [NW03a; NW03b; NW07], Kaashoek and Karger [KK03], Abraham et al. [AAA+03] and Fraigniaud and Gauron [FG03; FG06]) and fragment assembly for gene sequences (see Pevzner [Pev89], Pevzner and Tang [PT01] and Pevzner, Tang and Waterman [PTW01]). Note that these are just exemplary references. Actually, there is a much larger number of publications that relies on this kind of graphs.

The Kautz graphs were introduced in the articles of Kautz [Kau68; Kau71]. Similar to de Bruijn graphs, they can be used to represent strings of a certain length over a given alphabet. Again, two vertices are connected by an edge if the string representation of the source vertex can be transformed into the string representation of the target vertex by deleting the first symbol and adding a symbol at the end,

---

[4]Actually, they were already used implicitly by Flye Sainte-Marie [Fly94], see also de Bruijn [dBru75].

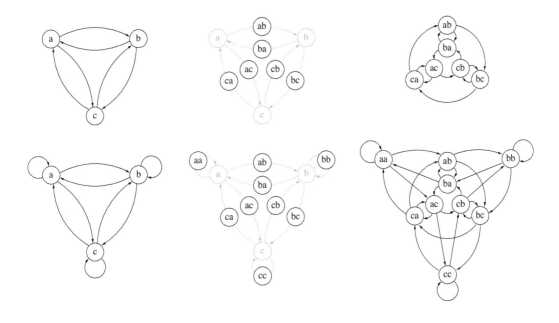

Figure 2.1: One step of the line digraph iteration for Kautz and de Bruijn graphs.

but in contrast to de Bruijn graphs, the strings are not allowed to contain the same symbol in consecutive positions. Therefore, Kautz graphs can be seen as a variant of de Bruijn graphs, where the loops are missing. An important field of application is the design of interconnection networks. For a given degree and diameter, Kautz graphs have a large number of vertices. Equivalently, a Kautz graph has a small diameter among the directed graphs with the same degree and the same number of vertices, see Fiol, Yebra and Alegre [FYA84]. Connectivity properties of graphs generated by line digraph iteration (like de Bruijn and Kautz graphs) were investigated by Du, Lyuu and Hsu [DLH93].

Recall that the edges of the line digraph $\mathcal{L}(G)$ represent exactly the walks of length 2 in the original digraph G while the vertices of $\mathcal{L}(G)$ correspond to the edges of G, i.e., to the walks of length 1. Based on this observation, one might suspect that there is a more general connection. It turned out that there is indeed a very close relationship between the k-step walks in a directed graph and the elements of its k-th iterated line digraph. At first, one might conjecture that repeating the operation leads to doubling the walk length with every iteration, but this is not true. As shown in the following theorem, the walk length is incremented by one in each iteration. Note that this connection was already observed by Cartwright and Gleason [CG66], and discovered again by Fiol, Yebra and Alegre [FYA84], Kwapisz [Kwa96], and Levine [Lev11]. Although it is quite fundamental, it seems to be largely unknown.

Theorem 2.2
The vertices of the k-th iterated directed line graph of a graph G are in one-to-one correspondence with the walks of length k in G. The edges of $\mathcal{L}_k(G)$ correspond to the walks of length $k+1$ in G.

Proof   For $\mathcal{L}_0(G) = G$, the statement is obvious. As already mentioned, we know for $\mathcal{L}_1(G) = \mathcal{L}(G)$ that the vertices correspond to the edges of G, i.e., to the walks of length 1. The edges in $\mathcal{L}(G)$ connect exactly the vertices of $\mathcal{L}(G)$ that are represented by original edges (s_1, t_1) and (s_2, t_2) with $t_1 = s_2$. Therefore, they represent exactly the walks of length 2: $(s_1, t_1, t_2) = (s_1, s_2, t_2)$. Now, the proof can be done by induction using the same principle for longer walks. Assume that the statement is true for $k = \ell$. Since the vertices of $\mathcal{L}_{\ell+1}(G) = \mathcal{L}(\mathcal{L}_\ell(G))$ correspond to the edges of $\mathcal{L}_\ell(G)$ and there is a bijection between the edges of $\mathcal{L}_\ell(G)$ and the walks of length $\ell + 1$ in G, there must be a bijection between the vertices of $\mathcal{L}_{\ell+1}(G)$ and the $(\ell + 1)$-walks in G. Consider two vertices x and y in $\mathcal{L}_{\ell+1}(G)$. They represent $(\ell+1)$-walks w_x and w_y and they correspond to edges (s_x, t_x) and (s_y, t_y) in $\mathcal{L}_\ell(G)$. Here s_x, t_x, s_y, and t_y correspond to ℓ-walks in G. The vertices x and y are connected by a directed edge (x, y) in $\mathcal{L}_{\ell+1}(G)$ if and only if $t_x = s_y$, i.e., the $(\ell+1)$-walks w_x and w_y overlap in the ℓ-walk $t_x = s_y$. Hence, each such edge (x, y) corresponds to a walk of length $k + 2$. Thus, the statement also holds for $k = \ell + 1$, and therefore for all $k \in \mathbb{N}$.

Corollary 2.1
For any directed graph G, the number of k-step walks in the ℓ-th iterated line digraph equals the number of $(k + \ell)$-step walks in G:

$$w_k\big(\mathcal{L}_\ell(G)\big) = w_{k+\ell}(G) \ .$$

The number of vertices in $\mathcal{L}_\ell(G)$ equals the number w_ℓ of ℓ-step walks in G while the number of edges in $\mathcal{L}_\ell(G)$ equals $w_{\ell+1}$.

This implies that the edge-connectivity of G equals the (vertex-)connectivity of $\mathcal{L}(G)$. As already mentioned, a further implication is that a line digraph $\mathcal{L}(G)$ has a Hamiltonian cycle if and only if its original graph G has an Eulerian cycle, see also Aigner [Aig67] and Kasteleyn [Kas67].

2.11   Other Applications

There is a vast number of other applications and it is impossible to discuss all of them in this book. This includes, for example, *additive number theory* (see Khintchine [Khi33] and Buchstab [Buc33]), *quantum probabilistic approaches* related to graph spectra (see Obata [Oba04], Hora and Obata [HO07] and Stevanović [Ste11]), close relations of graph eigenvalues and eigenvectors to the hubs and authorities of the *HITS algorithm* (see Benzi, Estrada and Klymko [BEK13]), *assortative mixing* in networks (see Newman [New02; New03], Van Mieghem et al. [VWG+10; VGS+10] and Liu, Trajanovski and Van Mieghem [LTV13]), *centrality and entropy* (see Benzi and Klymko [BK13], Estrada, de la Peña and Hatano [EdlPH14], Benzi [Ben14], Cao, Dehmer and Shi [CDS14], Li et al. [LQW+15] and Benzi and Klymko [BK15] and the discussions by Dehmer, Li and Shi [DLS15] and Dehmer and Mowshowitz [DM15]), as well as the *graph isomorphism problem* (see, e.g., Caporossi, Cvetković and Rowlinson [CCR14]).

Another field of application is concerned with the strong connection between matrix powers and certain integral forms called *iterated kernels*. We will investigate those relations within the last part of this book (see Chapter 9).

Surveys of applications in physics, chemistry, and other branches of science can be found in the books of Cvetković, Doob, Gutman and Torgašev [CDGT88] and Brualdi and Cvetković [BC09].

Chapter 3

Diagonalization and Spectral Decomposition

There exists, if I am not mistaken, an entire world which is the totality of mathematical truths, to which we have access only with our mind, just as a world of physical reality exists, the one like the other independent of ourselves, both of divine creation.

Charles Hermite

Although our primary interest —inequalities for the number of walks in graphs— is of an inherently enumerative nature, we will see that an algebraic view on the problem will turn out to be a valuable tool for deducing such combinatorial results. We will briefly review the eigenvalue approach for counting the number of walks in graphs using powers of adjacency matrices, and we will see how this can be generalized for other matrix classes.

3.1 Relevant Literature

For the first time, graph eigenvalues were discussed in a paper by Collatz and Sinogowitz [CS57].[1] In particular, they investigated relations between the largest eigenvalue of the adjacency matrix and the minimum, average, and maximum degree of the graph. Since adjacency matrices are nonnegative, the largest eigenvalue coincides with the spectral radius of the adjacency matrix (in [CS57] also called the *index* of the graph).

---

[1] Actually, the results had been obtained many years before the publication, see also the citation in the review of Seifert and Threlfall [ST48, p. 251]. Sinogowitz died already in September 1944.

Connections to the more general numbers of walks were investigated by Cvetković [Cve70c; Cve71], later also by Harary and Schwenk [HS79]. The early surveys by Godsil, Holton and McKay [GHM77] and Schwenk and Wilson [SW78] contain various results concerning graph eigenvalues.

Classic publications on spectral graph theory are the books by Biggs [Big74; Big93], Cvetković, Doob and Sachs [CDS79], Cvetković, Doob, Gutman and Torgašev [CDGT88], Godsil [God93], Chung [Chu97], and Cvetković, Rowlinson and Simić [CRS97]. Other books on algebraic graph theory were written by Godsil and Royle [GR01], Beineke and Wilson [BW05] and Knauer [Kna11]. Books on graphs and matrices that also contain the most important statements regarding the spectrum were written by Brualdi and Cvetković [BC09], Bapat [Bap10] and Brualdi [Bru11]. More recent overviews of spectral graph theory were written by Cvetković, Rowlinson and Simić [CRS09] and Spielman [Spi12]. Results on infinite graphs are treated in the survey of Mohar and Woess [MW89].

3.2   Similar and Diagonalizable Matrices

Let λ_i $(1 \leq i \leq n)$ denote the eigenvalues of the matrix A. Per definition, each of those values λ_i satisfies the equation $Av_i = \lambda_i v_i$ for some vector v_i, which is called an *eigenvector* corresponding to the *eigenvalue* λ_i.

A matrix B is called *similar* to matrix A if there is an invertible matrix S such that $B = S^{-1}AS$. Note that this similarity is an equivalence relation and that similar matrices have the same eigenvalues (counting multiplicities).

A matrix A is called *diagonalizable* if it is similar to a diagonal matrix D. The main diagonal entries of D are the eigenvalues λ_i of A. Therefore, we will also use the symbol Λ for this diagonal matrix.

Lemma 3.1
Suppose that A is a complex matrix that is diagonalizable by a similarity transformation $A = S^{-1}\Lambda S$ for an invertible matrix S and a diagonal matrix Λ. Then we have

$$\left(A^\ell\right)_{(i,j)} = \sum_{k=1}^{n} S^{-1}_{(i,k)} S_{(k,j)} \lambda_k^\ell \ .$$

Proof   Replacing A by $S^{-1}\Lambda S$, using associativity of matrix multiplication and cancelling out all the SS^{-1} factors (since they are equal to the identity matrix I), we obtain

$$A^\ell = \left(S^{-1}\Lambda S\right)^\ell = \underbrace{S^{-1}\Lambda \overbrace{SS^{-1}}^{=I} \Lambda S \ldots S^{-1}\Lambda \overbrace{SS^{-1}}^{=I} \Lambda S}_{\ell\text{-times } S^{-1}\Lambda S} = S^{-1}\Lambda^\ell S \ .$$

Note that Λ^ℓ is again a diagonal matrix containing the powers λ_i^ℓ of the eigenvalues of A on its main diagonal. The lemma follows now from the definition of matrix multiplication.

3.3 Hermitian and Real Symmetric Matrices

A well-known property of any self-adjoint matrix was shown by Hermite [Her55] and led to the name *Hermitian* matrix.

Theorem 3.1 (Hermite)
For every self-adjoint (Hermitian) matrix A, all eigenvalues $\lambda_1, \ldots, \lambda_n$ are real numbers.

In this case, we usually assume that the eigenvalues are indexed in nonincreasing order: $\lambda_1 \geq \ldots \geq \lambda_n$.

Cauchy proved that any Hermitian matrix is diagonalizable. More precisely, the *spectral theorem* for normal, Hermitian, and real symmetric matrices is the following statement (see, e.g., Horn and Johnson [HJ13]).

Theorem 3.2 (Spectral Theorem)
A matrix A is normal if and only if it can be diagonalized as $A = UDU^*$ by a unitary matrix U and a (complex) diagonal matrix D.
Moreover, A is Hermitian if and only if it can be diagonalized as $A = UDU^*$ by a unitary matrix U and a real diagonal matrix D.
Furthermore, A is a real symmetric matrix if and only if it can be diagonalized by a real orthogonal matrix U and a real diagonal matrix D such that $A = UDU^T$.

Notice that every unitary matrix U satisfies $U^{-1} = U^*$ and every orthogonal matrix U satisfies $U^{-1} = U^T$.

3.4 Adjacency Matrices and Number of Walks in Graphs

To get an impression of what we want to do, we briefly review the classical results for the special case of adjacency matrices of graphs, where U is a real orthogonal matrix. Here, we know by Lemma 1.1 that the number of walks of length ℓ from vertex v_i to vertex v_j is exactly the (i, j)-entry of the matrix $A^\ell = (U\Lambda U^T)^\ell = (U\Lambda U^{-1})^\ell = U\Lambda^\ell U^{-1} = U\Lambda^\ell U^T$. The number of walks from vertex v_i to vertex v_j is therefore as follows.

Lemma 3.2
In an undirected graph, the number of walks of length ℓ from vertex v_i to vertex v_j can be obtained from the spectral decomposition of the adjacency matrix $A = U\Lambda U^T$ as

$$w_\ell(v_i, v_j) = \sum_{k=1}^{n} u_{ik} u_{jk} \lambda_k^\ell \ .$$

The number of ℓ-step-walks starting at a given vertex v_i is

$$w_\ell(v_i) = \sum_{j=1}^{n}\sum_{k=1}^{n} u_{ik}u_{jk}\lambda_k^\ell = \sum_{k=1}^{n}\left(u_{ik}\lambda_k^\ell \sum_{j=1}^{n} u_{jk}\right) = \sum_{k=1}^{n} u_{ik}c_k(U)\lambda_k^\ell \ .$$

Then, the total number of walks of length ℓ can be calculated by

$$w_\ell = \sum_{i=1}^{n}\sum_{k=1}^{n} u_{ik}c_k(U)\lambda_k^\ell = \sum_{k=1}^{n} c_k(U)\lambda_k^\ell\left(\sum_{i=1}^{n} u_{ik}\right) = \sum_{k=1}^{n} c_k(U)^2\lambda_k^\ell \ .$$

For the sake of simplicity, we use $\hat{c}_j := (c_j(U))^2 = (\sum_{i=1}^{n} u_{ij})^2$ as abbreviations for the squared column sums of U.

Lemma 3.3
In an undirected graph, the total number of walks of length ℓ can be obtained from the spectral decomposition of the adjacency matrix $A = U\Lambda U^T$ as

$$w_\ell = \sum_{k=1}^{n} c_k(U)^2\lambda_k^\ell = \sum_{k=1}^{n} \hat{c}_k\lambda_k^\ell \ .$$

For the quadratic form of the real symmetric matrix A^ℓ, this corresponds to a reduction to the canonical form (to principal axes) by an orthogonal transformation (see Gantmacher [Gan60, Thm. 7, p. 309]). Note that the same applies to directed graphs that have a *normal* adjacency matrix. But in contrast to undirected graphs (where $\lambda_i \in \mathbb{R}$), the eigenvalues of directed graphs are complex numbers.

The total number of walks of length ℓ could also be written as (see Section 1.5)

$$w_\ell = \mathbf{1}_n^T A^\ell \mathbf{1}_n = \mathbf{1}_n^T\left(U\Lambda U^T\right)^\ell \mathbf{1}_n = \mathbf{1}_n^T U\Lambda^\ell U^T \mathbf{1}_n$$

or, using the inner product notation for real vectors, as

$$w_\ell = \left\langle \mathbf{1}_n, A^\ell \mathbf{1}_n \right\rangle = \left\langle \mathbf{1}_n, \left(U\Lambda U^T\right)^\ell \mathbf{1}_n \right\rangle = \left\langle \mathbf{1}_n, \left(U\Lambda^\ell U^T\right)\mathbf{1}_n \right\rangle \ .$$

In the next subsection, we will see that a similar result for the growth of the entry sum of matrix powers arises from the more general weighted sum

$$\mathrm{sum}_s\left(A^\ell\right) = s^T A^\ell s \ ,$$

using a certain scaling vector $s \in \mathbb{R}^n$. As already mentioned in Section 1.5, this introduces a possibility to select a set of start and end vertices of walks using the characteristic vector of this set.

From the diagonalization $U^T A U = \Lambda$, we can see that the k-th eigenvalue λ_k and the (unit) eigenvector $(u_{1k}\ldots u_{nk})^T$ are related by

$$\lambda_k = \sum_{(v_i,v_j)\in E} u_{ik}u_{jk} = \sum_{v_i\in V}\sum_{v_j\in V} a_{ij}u_{ik}u_{jk} = \sum_{i\in[n]}\sum_{j\in[n]} a_{ij}u_{ik}u_{jk} \ .$$

An even more general statement follows from $U^T A^\ell U = (U^T A U)^\ell = \Lambda^\ell$, namely

$$\lambda_k^\ell = \sum_{v_i\in V}\sum_{v_j\in V} w_\ell(v_i, v_j)u_{ik}u_{jk} = \sum_{i\in[n]}\sum_{j\in[n]} a_{i,j}^{[\ell]}u_{ik}u_{jk} \ .$$

By a similar argument, we obtain for $k \neq k'$ the equalities

$$0 = \sum_{(v_i, v_j) \in E} u_{ik} u_{jk'} = \sum_{i \in [n]} \sum_{j \in [n]} a_{ij} u_{ik} u_{jk'}$$

and

$$0 = \sum_{v_i \in V} \sum_{v_j \in V} w_\ell(v_i, v_j) u_{ik} u_{jk'} = \sum_{i \in [n]} \sum_{j \in [n]} a_{i,j}^{[\ell]} u_{ik} u_{jk'} \ .$$

The powers of eigenvalues are also directly related to the number of closed walks of length ℓ. This number corresponds to the sum of the entries $a_{i,i}^{[\ell]}$ of the main diagonal (i.e., the trace) of A^ℓ.

Lemma 3.4
For any graph, the total number of closed walks of length ℓ is

$$cl_\ell = \mathrm{tr}\left(A^\ell\right) = \sum_{k=1}^n \lambda_k^\ell \ .$$

Proof   If the eigenvalues of A are the complex numbers $\{\lambda_1, \ldots, \lambda_n\}$ then the eigenvalues of A^ℓ are the numbers $\{\lambda_1^\ell, \ldots, \lambda_n^\ell\}$. The lemma follows immediately from the fact that the trace of a matrix equals the sum of its eigenvalues.

Hence, we get $\sum_{k=1}^n \lambda_k = 0$ for loopless graphs. Otherwise, the sum of the eigenvalues equals the number of loops. For closed walks of length 2 in simple undirected graphs with m edges, we obtain $\sum_{k=1}^n \lambda_k^2 = 2m$. For bipartite graphs, we get even more restrictions since there are no closed walks of odd length. Then we have $\sum_{k=1}^n \lambda_k^{2k+1} = 0$ which goes with the fact that the spectrum of any bipartite graph is symmetric, i.e., for each eigenvalue λ_i, there is also an eigenvalue $\lambda_j = -\lambda_i$ with $j = n - i + 1$.

Various results to determine the number of walks for several graph constructions using spectral decompositions and generating functions can be found in the early papers and the thesis of Cvetković [Cve69; Cve70a; Cve70c; Cve70b; Cve71; Cve73]. For further discussions and references, see also Cvetković, Doob and Sachs [CDS79, Sect. 7.5].

3.5   Quadratic Forms for Diagonalizable Matrices

For the quadratic form of a normal matrix, that corresponds to a weighted entry sum of matrix powers (see Section 1.5), we have

$$\begin{aligned} \mathrm{sum}_s\left(A^\ell\right) &= s^* A^\ell s = \langle s, A^\ell s \rangle \\ &= s^*(U \Lambda U^*)^\ell s = s^* U \Lambda^\ell U^* s = \langle U^* s, \Lambda^\ell U^* s \rangle \ . \end{aligned}$$

For the sake of convenience, we assume that $c_s := U^* s$ and $c_{s,k} := (c_s)_k$. Thus, we have

$$\mathrm{sum}_s\left(A^\ell\right) = c_s^* \Lambda^\ell c_s = \langle c_s, \Lambda^\ell c_s \rangle \ .$$

From $A^\ell = (U\Lambda U^*)^\ell = U\Lambda^\ell U^*$, we know that

$$a_{i,j}^{[\ell]} = \left(A^\ell\right)_{(i,j)} = \sum_{k=1}^{n} u_{ik}\bar{u}_{jk}\lambda_k^\ell \ .$$

Since each entry in row i and column j will be weighted with the corresponding weights $\bar{s}_i$ and s_j, we define the following weighted version:

$$a_{i,j}^{[\ell,s]} = \bar{s}_i s_j \left(A^\ell\right)_{(i,j)} = \bar{s}_i s_j \sum_{k=1}^{n} u_{ik}\bar{u}_{jk}\lambda_k^\ell \ .$$

Now, we use the following generalized definitions for entry sums of matrix powers. For index $i \in [n]$, let $r_i^{[\ell,s]}$ denote the (row) sum of those weighted terms $a_{i,j}^{[\ell,s]}$ over all $j \in [n]$:

$$r_i^{[\ell,s]} = \sum_{j=1}^{n} a_{i,j}^{[\ell,s]} = \bar{s}_i \sum_{j=1}^{n} s_j \sum_{k=1}^{n} u_{ik}\bar{u}_{jk}\lambda_k^\ell = \bar{s}_i \sum_{k=1}^{n} \left(u_{ik}\lambda_k^\ell \sum_{j=1}^{n} s_j\bar{u}_{jk} \right) = \bar{s}_i \sum_{k=1}^{n} u_{ik}c_{s,k}\lambda_k^\ell \ .$$

Then, the total s-weighted sum of the entries of A^ℓ is

$$\mathrm{sum}_s\left(A^\ell\right) = \sum_{i=1}^{n} r_i^{[\ell,s]} = \sum_{i=1}^{n} \bar{s}_i \sum_{k=1}^{n} u_{ik}c_{s,k}\lambda_k^\ell = \sum_{k=1}^{n} \left(c_{s,k}\lambda_k^\ell \sum_{i=1}^{n} \bar{s}_i u_{ik} \right) = \sum_{k=1}^{n} c_{s,k}\bar{c}_{s,k}\lambda_k^\ell \ .$$

This means we have

$$\mathrm{sum}_s\left(A^\ell\right) = \langle c_s, c_s\Lambda^\ell \rangle = \langle d_s, d_s\Lambda^\ell \rangle \ ,$$

where d_s is the nonnegative vector containing the moduli of the entries of c_s, that is, $(d_s)_k := |(c_s)_k| = \sqrt{c_{s,k}\bar{c}_{s,k}}$. To simplify matters, we define $\hat{c}_{s,k} := c_{s,k}\bar{c}_{s,k} = |c_{s,k}|^2$. Thus, we have the following result.

Lemma 3.5
Suppose A is an $n \times n$ normal matrix and $s \in \mathbb{C}^n$ is a vector of complex numbers. Then, for all $\ell \in \mathbb{N}$, we have

$$\mathrm{sum}_s\left(A^\ell\right) = \sum_{k=1}^{n} \hat{c}_{s,k}\lambda_k^\ell$$

where $\hat{c}_{s,k} = c_{s,k}\overline{c_{s,k}} = (U^*s)_k\overline{(U^*s)_k}$ is defined by the spectral decomposition $A = U\Lambda U^*$ and $\lambda_i \in \mathbb{C}$ with $i \in [n]$ are the eigenvalues of A (that also define the diagonal matrix Λ).

Each $\hat{c}_{s,k}$ is a nonnegative real number since it is the product of a complex number and its complex conjugate. In principle, this corresponds to a reduction of the Hermitian form for the matrix A^ℓ to the canonical form by a unitary transformation of the variables (see Gantmacher [Gan60, Thm. 21, p. 337]). Note that by considering the weighted sum of A^0, i.e., $s^*A^0s = s^*s$, we obtain

$$\sum_{k=1}^{n} \hat{c}_{s,k} = \langle c_s, c_s \rangle = \langle s, s \rangle = |c_s|^2 = |s|^2 \ .$$

We can see that if the (real) eigenvalues of a Hermitian matrix satisfy $|\lambda_n| < \lambda_1$, the term $\sqrt[\ell]{\text{sum}_s\left(A^\ell\right)} = \sqrt[\ell]{\sum_{k=1}^n \hat{c}_{s,k}\lambda_k^\ell} = \sqrt[\ell]{\sum_{k=1}^n |c_{s,k}|^2\lambda_k^\ell}$ converges to λ_1 for $\ell \to \infty$, provided that $|c_{s,1}| > 0$. That is, the asymptotic growth of $\text{sum}_s\left(A^\ell\right)$ is governed by λ_1 in this case. In undirected multigraphs, the limit $\lim_{\ell \to \infty} \sqrt[\ell]{w_\ell/n} = \lim_{\ell \to \infty} \sqrt[\ell]{w_\ell}$ exists. This is sometimes called the *dynamic mean* of the vertex degrees and it is equal to the largest eigenvalue of the adjacency matrix (the *index* of the graph), see Cvetković [Cve71] and Cvetković, Doob and Sachs [CDS79, p. 46, p. 217]. A short discussion on such limits is also contained in the recent paper by Stevanović [Ste15]. For an arbitrary complex matrix A with spectral radius $\rho(A)$ and any (submultiplicative) matrix norm $\|.\|$, we have $\rho(A) = \lim_{p \to \infty} \|A^p\|^{1/p}$, see Theorem 8.3.

UNDIRECTED GRAPHS / HERMITIAN MATRICES

II

Chapter 4

General Results

As has been pointed out, beauty is in the eyes of the beholder. However, it is generally agreed that certain pieces of music, art, or mathematics are beautiful. There is an elegance to inequalities that makes them very attractive.

Richard Bellman

First of all, we will consider inequalities for products of certain walk numbers and products of quadratic forms of Hermitian matrices. Afterwards, we will investigate inequalities that involve powers of the number of walks and powers of quadratic forms.

4.1 Related Work for Products of Quadratic Forms

4.1.1 The Number of Walks in Graphs

4.1.1.1 The Results of Lagarias, Mazo, Shepp, and McKay

Lagarias, Mazo, Shepp and McKay [LMSM83] posed the following question: For which numbers $a, b \in \mathbb{N}$ is the inequality $w_a \cdot w_b \leq n \cdot w_{a+b}$ true for all graphs G? A little later, they proved the inequality for the case of an even sum $a + b$ [LMSM84]. Hence, it could be stated in the following way:

Theorem 4.1 (Lagarias et al.)
In any undirected graph on n vertices, the following inequality holds for all $a, b \in \mathbb{N}$:

$$w_{2a+b} \cdot w_b \leq n \cdot w_{2(a+b)} \ .$$

Furthermore, Lagarias et al. presented counterexamples for $w_a \cdot w_b \leq n \cdot w_{a+b}$ whenever $a + b$ is odd and $a, b \geq 1$. Nevertheless, they remarked without proof that, for any fixed graph, there is a constant c (depending on the graph) such that the inequality holds for all $a, b \geq c$.

4.1.1.2 The Results of Dress and Gutman

In the article of Gutman, Rücker and Rücker [GRR01], inequalities of the form

$$\frac{w_{2a-1}}{w_{2a-2}} \overset{?}{<} \frac{w_{2a}}{w_{2a-1}} \qquad \text{and} \qquad \frac{w_{2a}}{w_{2a-1}} \overset{?}{>} \frac{w_{2a+1}}{w_{2a}}$$

were discussed in the context of molecular graphs. The motivation behind those considerations was the simple fact that for odd walk lengths, the corresponding powers of the negative eigenvalues (in the sum resulting from the spectral theorem) yield negative terms, while those terms are positive for even walk lengths. The authors only noted that the inequalities are not true in general. While it was clear that there is equality for regular graphs anyway, they did not consider these statements any further because of two counterexamples. However, they did not notice that the counterexamples only apply to one of the two forms of the inequality and that the other one would be correct if the strict inequality is replaced by the corresponding relaxed version.

Dress and Gutman [DG03b] showed the following inequality:

Theorem 4.2 (Dress and Gutman)
In any undirected graph, the following inequality holds for all $a, b \in \mathbb{N}$:

$$w_{a+b}^2 \leq w_{2a} \cdot w_{2b} \ .$$

Note that the special case $b = a - 1$ corresponds to the relaxed form $\frac{w_{2a-1}}{w_{2a-2}} \leq \frac{w_{2a}}{w_{2a-1}}$ of the first inequality (from [GRR01]) above.

In a second paper, Dress and Gutman [DG03a] investigated the asymptotics of $w_{k+1} \cdot w_{k-1} - w_k^2$. They showed that the sign of this difference is constant for sufficiently large k. There, the interesting case occurs for even k (i.e., $k + 1$ and $k - 1$ are odd) since otherwise the difference is nonnegative anyway.

4.1.1.3 Further Results

Theorems 4.1 and 4.2 were generalized by our group (cf. Hemmecke et al. [HKM$^+$11; HKM$^+$12; TWK$^+$13]) to the following result, which we used to call the Sandwich Theorem.

Theorem 4.3
In any undirected graph, the following inequality holds for all $a, b, c \in \mathbb{N}$:

$$w_{2a+c} \cdot w_{2a+2b+c} \leq w_{2a} \cdot w_{2(a+b+c)} \ .$$

Later, we discovered the result of Marcus and Newman [MN62] for Hermitian matrices (see Theorem 4.4), that contains Theorem 4.3 as the special case for adjacency matrices of undirected graphs.

4.1.2 Real Symmetric, Positive-Semidefinite, and Hermitian Matrices

4.1.2.1 The Results of Marcus and Newman

The following theorems for entry sums of matrix powers were published by Marcus and Newman [MN62].

Theorem 4.4 (Marcus and Newman)
For every Hermitian matrix A and nonnegative integers $a, b, c \in \mathbb{N}$, the following inequality holds:

$$\operatorname{sum}\left(A^{2a+c}\right) \cdot \operatorname{sum}\left(A^{2a+2b+c}\right) \leq \operatorname{sum}\left(A^{2a}\right) \cdot \operatorname{sum}\left(A^{2(a+b+c)}\right) .$$

In the same paper, Marcus and Newman showed the following result for entry sums of a positive-semidefinite matrix.

Theorem 4.5 (Marcus and Newman)
For every positive-semidefinite symmetric matrix P and $k \in \mathbb{N}$, the following inequality holds:

$$\left(\operatorname{sum}\left(P^{k+1}\right)\right)^{2} \leq \operatorname{sum}\left(P^{k}\right) \cdot \operatorname{sum}\left(P^{k+2}\right) .$$

Both results are *unweighted* in the sense that the all-ones vector $\mathbf{1}_n$ can be used to calculate the simple entry sum of the matrix powers A^{ℓ} using the formula $\mathbf{1}_n^T A^{\ell} \mathbf{1}_n$. By contrast, the next results can be considered as an entry sum that is weighted using a certain scaling vector (as discussed in the introduction).

4.1.2.2 The Results of Lagarias, Mazo, Shepp, and McKay

Besides Theorem 4.1, Lagarias, Mazo, Shepp and McKay [LMSM84] proved the more general result

$$\left(v^{T} A^{r} v\right)\left(v^{T} A^{s} v\right) \leq \left(v^{T} v\right)\left(v^{T} A^{r+s} v\right)$$

for two cases of *weighted* entry sums. The first case requires an *even* sum $r + s$, real symmetric matrices A, and real vectors v.

Theorem 4.6 (Lagarias et al.)
For any real symmetric $n \times n$-matrix A, real vector $s \in \mathbb{R}^{n}$, and $a, b \in \mathbb{N}$, we have

$$\left(s^{T} A^{2a+b} s\right)\left(s^{T} A^{b} s\right) \leq \left(s^{T} s\right)\left(s^{T} A^{2(a+b)} s\right) .$$

The second case allows arbitrary natural numbers a and b, but requires positive-semidefinite matrices.

Theorem 4.7 (Lagarias et al.)
For any positive-semidefinite real symmetric $n \times n$-matrix P, real vector $s \in \mathbb{R}^n$, and $a, b \in \mathbb{N}$, we have

$$\left(s^T P^a s\right)\left(s^T P^b s\right) \leq \left(s^T s\right)\left(s^T P^{a+b} s\right) .$$

4.2   Generalizations for Products of Quadratic Forms

4.2.1   A Generalization of the Inequality by Dress and Gutman

For the proof of Theorem 4.2, Dress and Gutman [DG03b] simply applied the Cauchy-Schwarz inequality:

$$\left(\sum_{v \in V} w_a(v) \cdot w_b(v)\right)^2 \leq \left(\sum_{v \in V} w_a(v)^2\right)\left(\sum_{v \in V} w_b(v)^2\right) .$$

A similar approach is used in the book of Van Mieghem [Van11, p. 34] to obtain this result. Alternatively, Theorem 4.2 can be proven using the spectral decomposition, again using the Cauchy-Schwarz inequality. Then we have

$$\left[\sum_{i=1}^n \left(c_i \lambda_i^a \cdot c_i \lambda_i^b\right)\right]^2 \leq \left(\sum_{i=1}^n \left(c_i \lambda_i^a\right)^2\right)\left(\sum_{i=1}^n \left(c_i \lambda_i^b\right)^2\right) ,$$

which implies the result by the intermediate step

$$\left(\sum_{i=1}^n c_i^2 \lambda_i^{a+b}\right)^2 \leq \left(\sum_{i=1}^n c_i^2 \lambda_i^{2a}\right)\left(\sum_{i=1}^n c_i^2 \lambda_i^{2b}\right) .$$

Note that all involved numbers are real numbers.

Here we provide the following generalization.

Theorem 4.8
For all Hermitian matrices A, vectors $x, y \in \mathbb{C}^n$ and $a, b \in \mathbb{N}$, we have

$$\left|\operatorname{sum}_{x,y}\left(A^{a+b}\right)\right|^2 \leq \operatorname{sum}_x\left(A^{2a}\right) \cdot \operatorname{sum}_y\left(A^{2b}\right) .$$

Proof   The application of the Cauchy-Schwarz inequality to the vectors $x' = A^a x$ and $y' = A^b y$ yields (using $A^* = A$)

$$\left|x^* A^a A^b y\right|^2 \leq x^* A^a A^a x y^* A^b A^b y .$$

Corollary 4.1
In every undirected graph $G = (V, E)$, we have for all $X, Y \subseteq V$ and $a, b \in \mathbb{N}$

$$w_{a+b}(X, Y)^2 \leq w_{2a}(X, X) \cdot w_{2b}(Y, Y) .$$

Proof   The application of the Cauchy-Schwarz inequality to the vectors $x = A^a \chi(X)$ and $y = A^b \chi(Y)$ yields

$$\left| \chi(X)^T A^a A^b \chi(Y) \right|^2 \leq \chi(X)^T A^a A^a \chi(X) \chi(Y)^T A^b A^b \chi(Y) .$$

The special case $X = Y = V$, where X and Y consist of the whole vertex set, corresponds to Theorem 4.2. The next result corresponds to the special case where the sets X and Y consist of single vertices. It follows from the definition of the number of closed walks of length ℓ starting at a vertex v: $cl_\ell(v) = w_\ell(v, v)$.

Corollary 4.2
In every undirected graph $G = (V, E)$, we have for all $x, y \in V$ and $a, b \in \mathbb{N}$

$$w_{a+b}(x, y)^2 \leq cl_{2a}(x) \cdot cl_{2b}(y) .$$

4.2.2   The Method of Expectation of Random Variables

Lagarias, Mazo, Shepp and McKay [LMSM84] used the expectation of discrete random variables for the proof of Theorems 4.1 and 4.6. Their method can be extended in the following way. Suppose that A is a Hermitian matrix with spectral decomposition $A = U \Lambda U^*$. We consider the quadratic form

$$\mathrm{sum}_s\left(A^\ell\right) = s^* A^\ell s = s^* U \Lambda^\ell U^* s = \sum_{i=1}^{n} (U^* s)_i \overline{(U^* s)_i} \lambda_i^\ell = \sum_{i=1}^{n} \hat{c}_i \lambda_i^\ell$$

for a complex scaling vector $s \in \mathbb{C}^n$. Now, choose a $k \in \mathbb{N}$ and define a discrete random variable X taking the real values $\{\lambda_i\}$, $1 \leq i \leq n$, by

$$\Pr[X = \lambda_i] := \frac{(U^* s)_i \overline{(U^* s)_i} \lambda_i^{2k}}{\sum_{j=1}^{n} (U^* s)_j \overline{(U^* s)_j} \lambda_j^{2k}} = \frac{\hat{c}_i \lambda_i^{2k}}{\sum_{j=1}^{n} \hat{c}_j \lambda_j^{2k}} = \frac{\hat{c}_i \lambda_i^{2k}}{\mathrm{sum}_s\left(A^{2k}\right)} .$$

Note that it is necessary to use the even exponents $2k$ since the probabilities have to be nonnegative. In the special case where A is positive-semidefinite and thus all eigenvalues λ_i are nonnegative anyway, the factor of 2 can be left out and k can be a real number. Now we consider the expected value of X^r:

$$\mathrm{E}[X^r] = \frac{1}{\mathrm{sum}_s\left(A^{2k}\right)} \sum_{i=1}^{n} \hat{c}_i \lambda_i^{2k} \lambda_i^r = \frac{\mathrm{sum}_s\left(A^{2k+r}\right)}{\mathrm{sum}_s\left(A^{2k}\right)} .$$

In principle, this is just a weighted arithmetic mean where $\hat{c}_i \lambda_i^{2k}$ are the nonnegative weights of the terms λ_i^r. Similar to [LMSM84], we use the fact that

$$\mathrm{E}(X^a) \cdot \mathrm{E}(X^b) \leq \mathrm{E}(X^{a+b})$$

if $a + b$ is even or X is a nonnegative random variable. Let us remark that this also applies to all rational numbers a and b where λ_i^a and λ_i^b are defined and similarly ordered (or, in the case of a nonnegative random variable, for any $a, b \in \mathbb{R}$). Hence

$$\text{sum}_s\left(A^{2k+a}\right) \cdot \text{sum}_s\left(A^{2k+b}\right) \le \text{sum}_s\left(A^{2k}\right) \cdot \text{sum}_s\left(A^{2k+a+b}\right)$$

if $a + b$ is even (or under the more general assumptions above) or if A is positive-semidefinite. We will see a more elementary derivation of those results in the next two subsections.

4.2.3   The Complex-Weighted Sandwich Theorem

Theorem 4.9
For all Hermitian matrices A, nonnegative integers $a, b, c \in \mathbb{N}$, and scaling vectors $s \in \mathbb{C}^n$, the following inequality holds:

$$\text{sum}_s\left(A^{2a+c}\right) \cdot \text{sum}_s\left(A^{2a+2b+c}\right) \le \text{sum}_s\left(A^{2a}\right) \cdot \text{sum}_s\left(A^{2(a+b+c)}\right) \ .$$

Proof   We consider the difference of both sides of the inequality.

$$\text{sum}_s\left(A^{2a}\right) \cdot \text{sum}_s\left(A^{2(a+b+c)}\right) - \text{sum}_s\left(A^{2a+c}\right) \cdot \text{sum}_s\left(A^{2a+2b+c}\right)$$

$$= \sum_{i=1}^{n} \hat{c}_{s,i} \lambda_i^{2a} \sum_{j=1}^{n} \hat{c}_{s,j} \lambda_j^{2(a+b+c)} - \sum_{i=1}^{n} \hat{c}_{s,i} \lambda_i^{2a+c} \sum_{j=1}^{n} \hat{c}_{s,j} \lambda_j^{2a+2b+c}$$

$$= \sum_{i=1}^{n} \sum_{j=1}^{n} \hat{c}_{s,i} \hat{c}_{s,j} \left(\lambda_i^{2a} \lambda_j^{2(a+b+c)} - \lambda_i^{2a+c} \lambda_j^{2a+2b+c} \right)$$

For $i = j$, the difference is zero. For $i \ne j$, we combine the terms for $i < j$ and $i > j$:

$$= \sum_{i=1}^{n-1} \sum_{j=i+1}^{n} \hat{c}_{s,i} \hat{c}_{s,j} \left(\lambda_i^{2a} \lambda_j^{2(a+b+c)} - \lambda_i^{2a+c} \lambda_j^{2a+2b+c} + \lambda_j^{2a} \lambda_i^{2(a+b+c)} - \lambda_j^{2a+c} \lambda_i^{2a+2b+c} \right)$$

$$= \sum_{i=1}^{n-1} \sum_{j=i+1}^{n} \hat{c}_{s,i} \hat{c}_{s,j} \lambda_i^{2a} \lambda_j^{2a} \left(\lambda_j^{2(b+c)} - \lambda_i^c \lambda_j^{2b+c} + \lambda_i^{2(b+c)} - \lambda_j^c \lambda_i^{2b+c} \right)$$

$$= \sum_{i=1}^{n-1} \sum_{j=i+1}^{n} \hat{c}_{s,i} \hat{c}_{s,j} \lambda_i^{2a} \lambda_j^{2a} \left(\lambda_j^{2b+c} - \lambda_i^{2b+c} \right) \left(\lambda_j^c - \lambda_i^c \right).$$

Each term within the last line must be nonnegative, since $\hat{c}_{s,i}$, $\hat{c}_{s,j}$, λ_i^{2a}, and λ_j^{2a} are all nonnegative, and $(\lambda_j^{2b+c} - \lambda_i^{2b+c})$ and $(\lambda_j^c - \lambda_i^c)$ must have the same sign.

Theorem 4.6 by Lagarias et al. is a special case of Theorem 4.9 in which $a = 0$ and A is a real symmetric matrix.

Setting s to the characteristic vector $\chi(S)$ of an index subset $S \subseteq [n]$ gives a relation for the sum of entries restricted to the corresponding principal submatrix of the matrix power where we denote the sum of the corresponding matrix entries by sum $(A^k[S])$.

Corollary 4.3
For all Hermitian matrices A, nonnegative integers $a, b, c \in \mathbb{N}$, and subsets $S \subseteq [n]$, the following inequality holds:

$$\text{sum}\big(A^{2a+c}[S]\big) \cdot \text{sum}\big(A^{2a+2b+c}[S]\big) \leq \text{sum}\big(A^{2a}[S]\big) \cdot \text{sum}\big(A^{2(a+b+c)}[S]\big) .$$

By setting $S = [n]$ (i.e., $s = \mathbf{1}_n$ in Theorem 4.9), we obtain Theorem 4.4 (the result of Marcus and Newman [MN62]) as a special case. Notice that, in general, sum $(A^k[S])$ is different from sum $(A[S]^k)$, i.e., the entry sum of the k-th power of the principal submatrix. Applied to the adjacency matrix A of an undirected graph, sum $(A^k[S]) = w_k(S, S)$ is the number of walks of length k *starting* and *ending* at vertices of S, allowing all vertices of V as intermediate vertices. On the other hand, sum $(A[S]^k)$ is the number of walks where *all* vertices have to be in S, i.e., the number of walks of length k in the subgraph induced by S.

Corollary 4.4
For all undirected graphs $G = (V, E)$, any vertex subset $S \subseteq V$ and all $a, b, c \in \mathbb{N}$, we have

$$w_{2a+c}(S, S) \cdot w_{2a+2b+c}(S, S) \leq w_{2a}(S, S) \cdot w_{2(a+b+c)}(S, S) .$$

For $S = V$, this yields Theorem 4.3 (the inequality for the total number of walks). This in turn generalizes and unifies Theorem 4.1 (by Lagarias, Mazo, Shepp and McKay [LMSM84]) and Theorem 4.2 (by Dress and Gutman [DG03b]). In the case where S contains only one index $i \in [n]$ or one vertex $v \in V$, we obtain the following sandwich theorem for the main diagonal entries and the number of closed walks.

Corollary 4.5
For all Hermitian matrices A, all $a, b, c \in \mathbb{N}$, and $i \in [n]$, we have:

$$a_{i,i}^{[2a+c]} \cdot a_{i,i}^{[2a+2b+c]} \leq a_{i,i}^{[2a]} \cdot a_{i,i}^{[2(a+b+c)]}.$$

In undirected graphs, this corresponds to the following inequality for closed walks at $v \in V$:

$$cl_{2a+c}(v) \cdot cl_{2a+2b+c}(v) \leq cl_{2a}(v) \cdot cl_{2(a+b+c)}(v).$$

4.2.3.1 A Related Theorem by Taussky and Marcus

Recently, we found a theorem that is closely related to Theorem 4.9. Originally, it was proposed by Taussky [Tau60] as an open problem for positive-definite matrices in the "Advanced Problems and Solutions" section of *The American Mathematical Monthly.*

> "Let A, B be two positive definite hermitian matrices which can be transformed simultaneously by unitary transformations to diagonal forms of

similarly ordered numbers. Let x be any vector of complex numbers. Show that $\langle Ax, x \rangle \langle Bx, x \rangle \leq \langle ABx, x \rangle \langle x, x \rangle$, and discuss the case of equality."

The proof by Marcus [TM61][1] works for arbitrary (not only positive-definite) commutative Hermitian matrices A and B and is very similar to our proof of Theorem 4.9.

Theorem 4.10 (Taussky and Marcus)
We assume that A and B are commutative Hermitian matrices with eigenvalues $\lambda_1 \geq \ldots \geq \lambda_n$ and $\mu_1 \geq \ldots \geq \mu_n$ respectively. The conditions of the problem imply that there exists an orthonormal basis of common eigenvectors of A and B, $e_1, \ldots, e_n$, such that $Ae_i = \lambda_i e_i$, $Be_i = \mu_i e_i$, $i = 1, \ldots, n$. For any vector x let p_x and q_x be respectively the smallest and largest integers i for which $\sigma_i = |\langle x, e_i \rangle|^2 \neq 0$.
$\langle x, x \rangle \langle ABx, x \rangle - \langle Ax, x \rangle \langle Bx, x \rangle \geq 0$ *with equality if and only if either $\lambda_{p_x} = \lambda_{q_x}$ or $\mu_{p_x} = \mu_{q_x}$. If equality holds for an x for which $p_x = 1$ and $q_x = n$ then either A or B is a multiple of the identity.*

Note that since A and B are commuting Hermitian matrices, the product $AB = BA = B^* A^*$ is Hermitian, too. Thus, $\langle ABx, x \rangle$ is a real number.

Note also that two powers A^r and A^s of any matrix A are commuting. The eigenvalues of A^r and A^s are the corresponding powers λ_i^r and λ_i^s of the eigenvalues λ_i of A. We have $\lambda_i \in \mathbb{R}$ in case A is Hermitian. If r and s are either both odd or both even numbers then the new eigenvalues $\{\lambda_i^r\}$ and $\{\lambda_i^s\}$ are similarly ordered and Theorem 4.10 holds. Thus, Theorem 4.6 can be regarded as a special case of Theorem 4.10. Also Theorem 4.9 can be proven using Theorem 4.10 if we apply it to the matrices A^c, A^{2b+c} and the vector $x = sA^a$.

Similar inequalities can be obtained if A has only nonnegative eigenvalues (that is, if A is positive-semidefinite). If A has only nonpositive eigenvalues (that is, if A is negative-semidefinite), the direction of the inequality depends on whether the sum $r + s$ is an even or an odd number. These cases are handled in the next subsections.

4.2.4 Positive-Semidefinite Matrices

Now, we deduce a Sandwich Theorem for positive-semidefinite matrices that generalizes Theorem 4.5 (Marcus and Newman [MN62]) and Theorem 4.7 (Lagarias, Mazo, Shepp and McKay [LMSM84]). Remember that any positive real number x has a power $x^p \in \mathbb{R}$ for arbitrary real numbers p. Therefore, it is also possible to define a matrix power $A^p = U \Lambda^p U^*$ for positive-semidefinite matrices $A = U \Lambda U^*$ and $p \in \mathbb{R}$. Let sgn denote the sign function. Then we conclude the following.

---

[1]The theorem was also proven by A. F. Kaupe, Jr., R. F. Rinehart, J. E. Potter, and O. Taussky.

Theorem 4.11

For any positive-semidefinite matrix P, scaling vector $s \in \mathbb{C}^n$, and all $a, b, c \in \mathbb{R}$, the following inequalities hold.
If $\operatorname{sgn}(b) = \operatorname{sgn}(c)$ then

$$\operatorname{sum}_s\left(P^{a+b}\right) \cdot \operatorname{sum}_s\left(P^{a+c}\right) \leq \operatorname{sum}_s\left(P^a\right) \cdot \operatorname{sum}_s\left(P^{a+b+c}\right) .$$

If $\operatorname{sgn}(b) \neq \operatorname{sgn}(c)$ then

$$\operatorname{sum}_s\left(P^{a+b}\right) \cdot \operatorname{sum}_s\left(P^{a+c}\right) \geq \operatorname{sum}_s\left(P^a\right) \cdot \operatorname{sum}_s\left(P^{a+b+c}\right) .$$

Proof   The proof is essentially the same as for Theorem 4.9, except that squares of eigenvalues are not required since all eigenvalues are nonnegative in the case of positive-semidefinite matrices. Let $c_{s,i}$ be defined as before and again let $\hat{c}_{s,i} = c_{s,i}\bar{c}_{s,i}$. As in the proof of Theorem 4.9, we consider the difference of both sides of the inequality:

$$\sum_{i=1}^{n} \hat{c}_{s,i}\lambda_i^a \sum_{j=1}^{n} \hat{c}_{s,j}\lambda_j^{a+b+c} - \sum_{i=1}^{n} \hat{c}_{s,i}\lambda_i^{a+b} \sum_{j=1}^{n} \hat{c}_{s,j}\lambda_j^{a+c}$$

$$= \sum_{i=1}^{n}\sum_{j=1}^{n} \hat{c}_{s,i}\hat{c}_{s,j}\left(\lambda_i^a\lambda_j^{a+b+c} - \lambda_i^{a+b}\lambda_j^{a+c}\right)$$

$$= \sum_{i=1}^{n-1}\sum_{j=i+1}^{n} \hat{c}_{s,i}\hat{c}_{s,j}\left(\lambda_i^a\lambda_j^{a+b+c} - \lambda_i^{a+b}\lambda_j^{a+c} + \lambda_j^a\lambda_i^{a+b+c} - \lambda_j^{a+b}\lambda_i^{a+c}\right)$$

$$= \sum_{i=1}^{n-1}\sum_{j=i+1}^{n} \hat{c}_{s,i}\hat{c}_{s,j}\lambda_i^a\lambda_j^a\left(\lambda_j^{b+c} - \lambda_i^b\lambda_j^c + \lambda_i^{b+c} - \lambda_j^b\lambda_i^c\right)$$

$$= \sum_{i=1}^{n-1}\sum_{j=i+1}^{n} \hat{c}_{s,i}\hat{c}_{s,j}\lambda_i^a\lambda_j^a\left(\lambda_j^c - \lambda_i^c\right)\left(\lambda_j^b - \lambda_i^b\right) .$$

Again, $\hat{c}_{s,i}$ and $\hat{c}_{s,j}$ are nonnegative numbers. Furthermore, λ_i^a and λ_j^a are nonnegative, and $(\lambda_j^c - \lambda_i^c)$ and $(\lambda_j^b - \lambda_i^b)$ must have the same sign if b and c have the same sign since $\lambda_i, \lambda_j \geq 0$. (The sequences $\lambda_1^b, \ldots, \lambda_n^b$ and $\lambda_1^c, \ldots, \lambda_n^c$ are similarly ordered.) Therefore, each term within the last line must be nonnegative. If b and c have different signs, then $(\lambda_j^c - \lambda_i^c)$ and $(\lambda_j^b - \lambda_i^b)$ have opposing signs, too. (The sequences $\lambda_1^b, \ldots, \lambda_n^b$ and $\lambda_1^c, \ldots, \lambda_n^c$ are conversely ordered.) Therefore, the sum is nonpositive.

As before, setting s to the characteristic vector of an index subset $S \subseteq [n]$ yields the following result for principal submatrices.

Corollary 4.6

For all positive-semidefinite matrices P, all $a, b, c \in \mathbb{N}$, and all subsets $S \subseteq [n]$, we have

$$\operatorname{sum}\left(P^{a+b}[S]\right) \cdot \operatorname{sum}\left(P^{a+c}[S]\right) \leq \operatorname{sum}\left(P^a[S]\right) \cdot \operatorname{sum}\left(P^{a+b+c}[S]\right) .$$

4.2.5 Negative-Semidefinite Matrices

The missing cases for negative-semidefinite matrices can be deduced from the last line of the previous proof.

Theorem 4.12
For any negative-semidefinite matrix N, scaling vector $s \in \mathbb{C}^n$, and all $a, b, c \in \mathbb{N}$, the following inequalities hold.
If $b + c$ is even, we have

$$\operatorname{sum}_s\left(N^{a+b}\right) \cdot \operatorname{sum}_s\left(N^{a+c}\right) \leq \operatorname{sum}_s\left(N^a\right) \cdot \operatorname{sum}_s\left(N^{a+b+c}\right) \ .$$

If $b + c$ is odd, we have

$$\operatorname{sum}_s\left(N^{a+b}\right) \cdot \operatorname{sum}_s\left(N^{a+c}\right) \geq \operatorname{sum}_s\left(N^a\right) \cdot \operatorname{sum}_s\left(N^{a+b+c}\right) \ .$$

Proof   We refer to the same transformation as in the previous proof. Again, $\hat{c}_{s,i}$, $\hat{c}_{s,j}$, and $\lambda_i^a \lambda_j^a$ are nonnegative. Now, the sequences $\lambda_1^b, \ldots, \lambda_n^b$ and $\lambda_1^c, \ldots, \lambda_n^c$ are similarly ordered if both numbers b and c are even, or both are odd, i.e., $b + c$ is even (since $\lambda_i \leq 0$). They are conversely ordered if $b + c$ is odd.

Note that the statement could be slighly extended to the rational powers of negative eigenvalues (and hence negative-semidefinite matrices) that are defined.

Again, setting s to the characteristic vector of an index subset $S \subseteq [n]$ yields the following result for principal submatrices.

Corollary 4.7
For all negative-semidefinite matrices N, all $a, b, c \in \mathbb{N}$, and all subsets $S \subseteq [n]$, the following inequalities hold.
If $b + c$ is even, we have

$$\operatorname{sum}\left(N^{a+b}[S]\right) \cdot \operatorname{sum}\left(N^{a+c}[S]\right) \leq \operatorname{sum}\left(N^a[S]\right) \cdot \operatorname{sum}\left(N^{a+b+c}[S]\right) \ .$$

If $b + c$ is odd, we have

$$\operatorname{sum}\left(N^{a+b}[S]\right) \cdot \operatorname{sum}\left(N^{a+c}[S]\right) \geq \operatorname{sum}\left(N^a[S]\right) \cdot \operatorname{sum}\left(N^{a+b+c}[S]\right) \ .$$

4.2.6 The Density Implication

Graph Density.

For a graph G having $n \geq 2$ vertices and m edges, the *density* $\rho(G)$ is defined as the fraction of present edges:

$$\rho(G) := \frac{m}{\binom{n}{2}} = \frac{2m}{n(n-1)} \ .$$

Notice that this definition was made for undirected graphs without loops and without parallel edges. For loop-free directed graphs having n vertices and m edges, the definition would be $\rho(G) := \frac{m}{n(n-1)}$.

Accordingly, a generalized *density of order* ℓ can be defined using the total number of ℓ-step walks (see Kosub [Kos05, p. 132]):

$$\rho_\ell(G) := \frac{w_\ell}{n(n-1)^\ell} \ .$$

If loops shall be allowed, then the term $(n-1)$ in the denominator should be replaced by n in all the definitions above. Mostly, we will use ρ and ρ_ℓ instead of $\rho(G)$ and $\rho_\ell(G)$ if G is clear from the context. As expected, we have $\rho_0 = 1$ and $\rho_1 = \rho$.

Theorem 4.3 directly implies the following inequality:

$$\frac{w_{2a+c} \cdot w_{2a+2b+c}}{[n(n-1)^{2a+c}] \cdot [n(n-1)^{2a+2b+c}]} \leq \frac{w_{2a} \cdot w_{2(a+b+c)}}{[n(n-1)^{2a}] \cdot [n(n-1)^{2(a+b+c)}]} \ .$$

Corollary 4.8
In any undirected graph, the following inequality holds for all $a, b, c \in \mathbb{N}$:

$$\rho_{2a+c} \cdot \rho_{2a+2b+c} \leq \rho_{2a} \cdot \rho_{2(a+b+c)} \ .$$

Note that the same inequality also holds for the case, where loops are allowed.

The sandwich theorem for the total number of walks in undirected graphs can also be used to show the following bound for the average degree (using $a = 0$ and $c = 1$):

$$\bar{d} = \frac{w_1}{w_0} \leq \frac{w_{2b+2}}{w_{2b+1}} \ .$$

Matrix Density.

In the case of general symmetric or Hermitian matrices, we get the following statement regarding a certain measure of density.

$$\frac{\text{sum}(A)}{n} = \frac{\text{sum}(A^1)}{\text{sum}(A^0)} \leq \frac{\text{sum}(A^{2k})}{\text{sum}(A^{2k-1})}$$

That means, the average row sum obeys the inequality

$$\frac{1}{n} \sum_{i=1}^n r_i \leq \frac{\text{sum}(A^{2k})}{\text{sum}(A^{2k-1})}$$

and the arithmetic mean $\bar{a} := \text{sum}(A)/n^2$ of all entries of A observes the inequality

$$\bar{a} \leq \frac{1}{n} \cdot \frac{\text{sum}(A^{2k})}{\text{sum}(A^{2k-1})} \ .$$

4.3 Related Work for Powers of Quadratic Forms

In the following, we briefly review known results where powers of quadratic forms are involved. Again, we start with results concerning the number of walks in undirected graphs.

4.3.1 The Results of Erdős, Simonovits, and Godsil

In their paper on compactness results in extremal graph theory, Erdős and Simonovits [ES82][2] considered the following inequality.

Theorem 4.13 (Erdős and Simonovits)
For all undirected graphs and $k \in \mathbb{N}$, we have

$$w_k \geq n \cdot \overline{d}^k \ .$$

This generalized some previous work by Faudree and McKay,[3] who had proved the theorem for $k = 3^p$, $p \in \mathbb{N}$ (see [ES82]). It was noticed by Godsil that Theorem 4.13 could be proven using the results of Mulholland and Smith [MS59], Blakley and Roy [BR65], and London [Lon66a]. Note that also the articles by Scheuer and Mandel [SM59] and Blakley and Dixon [BD66] imply Theorem 4.13. All these results will be discussed later.

Recall that $\overline{d} = 2m/n$. Now, Theorem 4.13 can be interpreted in several ways, for example as a comparison of averages: $\frac{w_k}{n} \geq \overline{d}^k$. The average number of k-step walks per vertex is greater than or equal to the k-th power of the average degree or the average number of 1-step walks per vertex. The theorem could also be formulated in terms of the number of edges and the average degree:

$$w_k \geq 2m \cdot \overline{d}^{k-1} \ .$$

Using the simple facts $w_1 = 2m$ and $w_0 = n$, we can rewrite the inequality in another way that only uses walk numbers:

$$w_k \geq w_0 \left(\frac{w_1}{w_0} \right)^k \qquad \text{or} \qquad w_1^k \leq w_0^{k-1} \cdot w_k \ .$$

For trees, we have $m = n - 1$. Then Theorem 4.13 implies

$$\frac{w_k}{n} \geq \left(\frac{2(n-1)}{n} \right)^k = \left(2 - \frac{2}{n} \right)^k = 2^k \left(1 - \frac{1}{n} \right)^k \ .$$

Actually, Erdős and Simonovits [ES82] proved the following more general result.

---

[2]Actually, also C. D. Godsil contributed to this paper, see the comments in [ES82].
[3]R. J. Faudree and B. D. McKay: The number of walks of length a power of three in a graph. Preprint (1980).

Theorem 4.14 (Erdős and Simonovits)
Let

$$f(x) = x^k + a_1 x^{k-1} + a_2 x^{k-2} + \ldots + a_k \ ,$$

for even k and

$$F(G) = \frac{w_k}{n} + a_1 \frac{w_{k-1}}{n} + \ldots + a_k \frac{w_0}{n} \ .$$

If f is a convex increasing function for $x \geq x_0$ and m denotes its minimum in $(-\infty, \infty)$ then for $\overline{d} \geq x_0$

$$F(G) \geq f(\overline{d}) + m - f(x_0) \ .$$

As noted in [ES82], an important case of the theorem is the special case where f has k real roots the maximum of which is x_0. (Then $f(x_0) = 0$.)

Additionally, Godsil obtained the following result [ES82].

Theorem 4.15 (Godsil)
For all undirected graphs and $k, t \in \mathbb{N}$, we have

$$\left(\frac{w_{2k}}{n}\right)^t \geq \left(\frac{w_t}{n}\right)^{2k} \qquad for \qquad 2k \geq t \ .$$

This can be written as

$$w_t^{2k} \leq w_0^{2k-t} \cdot w_{2k}^t \ .$$

Further, Erdős and Simonovits [ES82] made the following conjecture.

Conjecture 4.1 (Erdős and Simonovits)
For all undirected graphs, we have

$$\left(\frac{w_k}{n}\right)^t \geq \left(\frac{w_t}{n}\right)^k \qquad for \qquad k \geq t \ and \ t, k \ are \ odd \ .$$

Note that this problem was already investigated by Blakley and Dixon [BD66]. They could not give a general proof, but sufficient conditions with huge constants for the inequality to hold.

In a more recent paper, Alon, Hoory and Linial [AHL02] proved a theorem for nonreturning walks that is very similar to Theorem 4.13.

Theorem 4.16 (Alon, Hoory and Linial)
Suppose G is a graph of average degree $\overline{d} \geq 2$. Then

$$\frac{\nu_k}{n} \geq \overline{d}(\overline{d} - 1)^{k-1} \ .$$

If $r \in \mathbb{N}$ then equality holds if and only if G is regular.

4.3.2 The Results of Ilić and Stevanović

For the Zagreb indices $M_1 = \sum_{v \in V} d_v^2$ and $M_2 = \sum_{\{x,y\} \in E} d_x d_y$, the following inequalities were proposed by Ilić and Stevanović [IS09].

Theorem 4.17 (Ilić and Stevanović)
For all undirected graphs, we have

$$\frac{M_1}{n} \geq \frac{4m^2}{n^2} \ .$$

Since $M_1 = w_2$ and $4m^2 = (2m)^2 = w_1^2$, Theorem 4.17 is equivalent to

$$w_1^2 \leq w_0 w_2 \ .$$

Thus, it is a special case of Theorems 4.1, 4.2, 4.4, 4.13, and 4.15.

Theorem 4.18 (Ilić and Stevanović)
For all undirected graphs, we have

$$\frac{M_2}{m} \geq \frac{4m^2}{n^2} \ .$$

By $M_2 = w_3/2$ and $w_1 = 2m$, we see that Theorem 4.18 is equivalent to

$$w_1^3 \leq w_0^2 w_3 \ .$$

This is a special case of Theorem 4.13.

4.3.3 The Results of Scheuer and Mandel, Mulholland and Smith, Blakley and Roy, Blakley and Dixon, and London

Scheuer and Mandel [SM59] showed the following result.

Theorem 4.19 (Scheuer and Mandel)
For any nonnegative symmetric matrix A and any unit vector u with positive elements, the inequality

$$u^T A^r u \geq (u^T A u)^r$$

holds for all integers $r > 0$.

As noted in the paper by Erdős and Simonovits [ES82], the following theorem of Mulholland and Smith [MS59; MS60][4] can be used to prove Theorem 4.13.

---

[4]According to [MS59], an alternative proof was provided by M. M. Crum and R. Scheuer.

Theorem 4.20 (Mulholland and Smith)
Suppose A is a nonzero nonnegative symmetric matrix and v is a nonzero nonnegative vector of the same dimension. Then, for any positive integer k,

$$(v^T A^k v)(v^T v)^{k-1} \geq (v^T A v)^k$$

with equality if and only if v is an eigenvector of A.

Note that this inequality also holds if A is the zero matrix. Furthermore, it holds if $k = 0$ and v is not the zero vector, or if $k \geq 2$ in case that v is the zero vector.

In the proof of Theorem 4.20, Mulholland and Smith [MS59] also mentioned the following "standard result" for *real* symmetric positive-semidefinite matrices and arbitrary *real* vectors, i.e., A and v might contain negative entries.

Theorem 4.21
Suppose P is a nonzero real symmetric matrix with nonnegative eigenvalues (i.e., P is positive-semidefinite) and v is a nonzero real vector of the same dimension. Then, for any positive integer k,

$$(v^T P^k v)(v^T v)^{k-1} \geq (v^T P v)^k$$

with equality if and only if v is an eigenvector of P.

The inequality corresponds to a special case of Theorem 4.31.

Further, Mulholland and Smith remark that the inequality

$$\left(\text{sum}_s(A^m)\right)^{n-h} \leq \left(\text{sum}_s(A^n)\right)^{m-h} \left(\text{sum}_s(A^h)\right)^{n-m}$$

is also true even if some $\lambda_i < 0$, if n and h are even (see also Theorem 4.30).

Theorem 4.20 appeared in slightly different form in a later paper by Blakley and Roy [BR65].

Theorem 4.22 (Blakley and Roy)
If A is a nonnegative symmetric matrix, u is a unit nonnegative vector and k is a positive integer then

$$\langle u, Au \rangle^k \leq \langle u, A^k u \rangle \ .$$

If $k > 1$ equality holds if and only if u is an eigenvector of A or $\langle u, A^k u \rangle = 0$.

Wimmer [Wim85] showed that for the case of equality $\langle u, Au \rangle^k = \langle u, A^k u \rangle$, $k > 1$, the condition $\langle u, A^k u \rangle = 0$ is redundant, i.e., for $k > 1$ equality holds if and only if u is an eigenvector of A. With this modification, Theorem 4.22 is exactly the same as Theorem 4.20, except for the fact that normalized vectors are calculated implicitly within the inequality of Theorem 4.20 and a unit vector is assumed in advance in Theorem 4.22. A generalization of Theorems 4.20 and 4.22 to a certain variation of powers of nonsymmetric matrices will be discussed in a later part of this book (see Theorem 6.7).

Blakley and Dixon [BD66] investigated the more general inequality

$$\langle v, A^j v \rangle^{j+k} \overset{?}{\le} \langle v, v \rangle^k \langle v, A^{j+k} v \rangle^j \ .$$

for positive integers j, k, a symmetric real $n \times n$-matrix A, and a real n-dimensional vector v. They made the following observations.

Theorem 4.23 (Blakley and Dixon)
For any positive-semidefinite matrix P, the inequality

$$\langle v, P^j v \rangle^{j+k} \le \langle v, v \rangle^k \langle v, P^{j+k} v \rangle^j$$

holds for each real vector $v \in \mathbb{R}^n$ and positive integers j and k.

As noted in [BD66], this can be proven using Hölder's inequality. The special case for $j = 1$ corresponds to Theorem 4.21. A consequence of the positive-semidefinite case is the following special case.

Corollary 4.9 (Blakley and Dixon)
If $j + k$ is even for positive integers j and k then

$$\langle v, A^j v \rangle^{j+k} \le \langle v, v \rangle^k \langle v, A^{j+k} v \rangle^j$$

holds for all real vectors $v \in \mathbb{R}^n$ and all symmetric real matrices A.

On the other hand, it is also known that the inequality does not hold in full generality.

Theorem 4.24 (Blakley and Dixon)
For positive integers j and k where j is even and k is odd, there is a nonnegative symmetric matrix A and a nonnegative vector v with

$$\langle v, A^j v \rangle^{j+k} \not\le \langle v, v \rangle^k \langle v, A^{j+k} v \rangle^j \ .$$

As a counterexample, Blakley and Dixon mention the case

$$v = \begin{pmatrix} 0 \\ 1 \\ 2 \end{pmatrix} \qquad \text{and} \qquad A = \begin{pmatrix} 0 & 0 & 2 \\ 0 & 1 & 0 \\ 2 & 0 & 0 \end{pmatrix} \ .$$

It is not clear whether the inequality holds in the case where j is odd and k is even. For the special case of the number of walks in graphs, i.e., the entry sum of powers in adjacency matrices, this corresponds to Conjecture 4.1.

In the special case $j = 1$, the inequality holds for nonnegative vectors v and nonnegative matrices A. Then we have

$$\langle v, Av \rangle^{k+1} \le \langle v, v \rangle^k \langle v, A^{k+1} v \rangle \ .$$

If A has a nonzero entry, v is not on the boundary of the half cone consisting of all nonnegative n-dimensional vectors and v is not an eigenvector of A, then the inequality is strict. As discussed before, this is essentially the same statement as Theorems 4.20 and 4.22 (without prior normalization of the vector v). However, Blakley and Dixon presented an alternative proof of the statement to introduce another method using Fréchet differentiation.

4.3.4   The Results of Hyyrö, Merikoski, and Virtanen

The following results are due to Hyyrö, Merikoski and Virtanen [HMV86].

Theorem 4.25 (Hyyrö et al.)
Assume that A is a real symmetric matrix and let $p_1 \geq \ldots \geq p_h$, $q_1 \geq \ldots \geq q_h \in \mathbb{N}$, such that $p_1 \geq \ldots \geq p_h$ are even numbers and the p_i's majorize the q_i's.
Then
$$\mathrm{sum}\left(A^{p_1}\right) \cdot \ldots \cdot \mathrm{sum}\left(A^{p_h}\right) \geq \mathrm{sum}\left(A^{q_1}\right) \cdot \ldots \cdot \mathrm{sum}\left(A^{q_h}\right) \ .$$

Theorem 4.26 (Hyyrö et al.)
Assume that A is a real symmetric matrix with $\mathrm{sum}\left(A^{2\ell}\right) \neq 0$ and $\mathrm{sum}\left(A^{2s}\right) \neq 0$, k is even, $\ell \geq s$, and $k + 2\ell \geq r + 2s$.
Then
$$\sqrt[k]{\frac{\mathrm{sum}\left(A^{k+2\ell}\right)}{\mathrm{sum}\left(A^{2\ell}\right)}} \geq \sqrt[r]{\frac{\mathrm{sum}\left(A^{r+2s}\right)}{\mathrm{sum}\left(A^{2s}\right)}} \ .$$

For the case where A is nonnegative and r divides k, but k does not have to be an even number, Hyyrö et al. also obtained the following by simply applying Theorem 4.20 to the matrix $B = A^r$.

Theorem 4.27 (Hyyrö et al.)
If A is a nonnegative symmetric $n \times n$-matrix and $m, r \in \mathbb{N}$, then
$$\sqrt[mr]{\frac{\mathrm{sum}\left(A^{mr}\right)}{n}} \geq \sqrt[r]{\frac{\mathrm{sum}\left(A^{r}\right)}{n}} \ .$$

Note that all results from [HMV86] deal with *unweighted* entry sums of the matrix powers (i.e., without scaling the rows and columns).

4.4   Generalizations for Powers of Quadratic Forms

4.4.1   More General Results for Walk Numbers

While all of our inequalities are related to powers, namely the entry sum for powers of matrices, we now turn to inequalities that involve powers of those sums.

Note that Theorems 4.20 and 4.22 can be written in the following way:

$$\left(\frac{v^T A v}{v^T v}\right)^k \leq \left(\frac{v^T A^k v}{v^T v}\right) \qquad \text{or} \qquad \left(\frac{\text{sum}_v(A)}{\|v\|^2}\right)^k \leq \left(\frac{\text{sum}_v(A^k)}{\|v\|^2}\right) \, ,$$

which corresponds to normalized weighted sums.

As a generalization of Theorem 4.2 (Dress and Gutman [DG03b]), Theorem 4.13 (Erdős and Simonovits [ES82]), Theorems 4.17 and 4.18 (Ilić and Stevanović [IS09]), and Theorem 4.27 (Hyyrö, Merikoski and Virtanen [HMV86]), we propose the following theorem.

Theorem 4.28

For every Hermitian $n \times n$ matrix A, any scaling vector $s \in \mathbb{C}^n$, and all $k, \ell, p \in \mathbb{N}$ such that A^p and $A^\ell s$ are nonnegative, the following inequality holds if $k \geq 2$ or $\text{sum}_s\left(A^{2\ell}\right) > 0$:

$$\left(\text{sum}_s\left(A^{2\ell+p}\right)\right)^k \leq \left(\text{sum}_s\left(A^{2\ell}\right)\right)^{k-1} \cdot \text{sum}_s\left(A^{2\ell+pk}\right).$$

For all matrices with $\text{sum}_s\left(A^{2\ell}\right) > 0$, this is equivalent to

$$\left(\frac{\text{sum}_s\left(A^{2\ell+p}\right)}{\text{sum}_s\left(A^{2\ell}\right)}\right)^k \leq \frac{\text{sum}_s\left(A^{2\ell+pk}\right)}{\text{sum}_s\left(A^{2\ell}\right)} \quad and \quad \left(\frac{\text{sum}_s\left(A^{2\ell+p}\right)}{\text{sum}_s\left(A^{2\ell}\right)}\right)^{k-1} \leq \frac{\text{sum}_s\left(A^{2\ell+pk}\right)}{\text{sum}_s\left(A^{2\ell+p}\right)}.$$

Proof Assume that the preconditions of the theorem are fulfilled, then we can apply Theorem 4.20 to the nonnegative symmetric matrix $B := A^p$ and the nonnegative vector $v := A^\ell s$. Note that v is nonnegative and therefore $(A^\ell s)^T = (A^\ell s)^* = s^*(A^*)^\ell$. Since A is Hermitian we have $A^* = A$ and therefore

$$\left[\left(A^\ell s\right)^T A^p \left(A^\ell s\right)\right]^k \leq \left[\left(A^\ell s\right)^T \left(A^p\right)^k \left(A^\ell s\right)\right] \cdot \left[\left(A^\ell s\right)^T \left(A^\ell s\right)\right]^{k-1}$$

$$\left[s^* A^\ell A^p A^\ell s\right]^k \leq \left[s^* A^\ell A^{pk} A^\ell s\right] \cdot \left[s^* A^\ell A^\ell s\right]^{k-1} .$$

Summing up the exponents finishes the proof.

Theorems 4.20 and 4.22 correspond to the special case $\ell = 0$ and $p = 1$ ($\|s\|^2 = s^T s = \text{sum}_s(A^0)$). Theorem 4.27 corresponds to the unweighted case $s = \mathbf{1}_n$ ($\mathbf{1}_n^T \mathbf{1}_n = n$) for $\ell = 0$.

Corollary 4.10

For each nonnegative real symmetric matrix A, subset $S \subseteq [n]$, and $k, \ell, p \in \mathbb{N}$, the following inequality holds if $k \geq 2$ or $\text{sum}\left(A^{2\ell}[S]\right) > 0$:

$$\left(\text{sum}\left(A^{2\ell+p}[S]\right)\right)^k \leq \left(\text{sum}\left(A^{2\ell}[S]\right)\right)^{k-1} \cdot \text{sum}\left(A^{2\ell+pk}[S]\right) .$$

If the matrix is the adjacency matrix of a graph $G = (V, E)$ and s is the characteristic vector of a vertex subset $S \subseteq V$, then we obtain the following result for walks that start and end at vertices of S (where the intermediate vertices may also come from $V \setminus S$).

Corollary 4.11
For every graph $G = (V, E)$, vertex subset $S \subseteq V$, and $k, \ell, p \in \mathbb{N}$, the following inequality holds if $k \geq 2$ or $w_{2\ell}(S, S) > 0$:

$$w_{2\ell+p}(S, S)^k \leq w_{2\ell}(S, S)^{k-1} \cdot w_{2\ell+pk}(S, S).$$

For all graphs with $w_{2\ell}(S, S) > 0$, this is equivalent to

$$\left(\frac{w_{2\ell+p}(S, S)}{w_{2\ell}(S, S)} \right)^k \leq \frac{w_{2\ell+pk}(S, S)}{w_{2\ell}(S, S)} \quad and \quad \left(\frac{w_{2\ell+p}(S, S)}{w_{2\ell}(S, S)} \right)^{k-1} \leq \frac{w_{2\ell+pk}(S, S)}{w_{2\ell+p}(S, S)}.$$

If the subset S includes all of the vertices, then we get the following result.

Corollary 4.12
For every graph and all $k, \ell, p \in \mathbb{N}$ such that $k \geq 2$ or $w_{2\ell} > 0$, we have

$$w_{2\ell+p}^k \leq w_{2\ell}^{k-1} \cdot w_{2\ell+pk} \ .$$

For all graphs with $w_{2\ell} > 0$ (in particular for graphs with at least one edge), this is equivalent to

$$\left(\frac{w_{2\ell+p}}{w_{2\ell}} \right)^k \leq \frac{w_{2\ell+pk}}{w_{2\ell}} \quad and \quad \left(\frac{w_{2\ell+p}}{w_{2\ell}} \right)^{k-1} \leq \frac{w_{2\ell+pk}}{w_{2\ell+p}} \ .$$

Corollary 4.12 is a generalization of Theorem 4.13 (the inequality of Erdős and Simonovits).

Setting $k = 2$ leads to Theorem 4.2, $w_{2\ell+p}^2 \leq w_{2\ell+2p} \cdot w_{2\ell}$, i.e., the inequality by Dress and Gutman [DG03b]. Furthermore, Corollary 4.12 is a generalization of Theorems 4.17 and 4.18, i.e., the two inequalities that were proposed by Ilić and Stevanović [IS09] for the Zagreb indices.

If the chosen subset S contains only a single vertex v, then we obtain the following statement about the number of closed walks starting at the given vertex v.

Corollary 4.13
For every graph $G = (V, E)$, every vertex $v \in V$, and all $k, \ell, p \in \mathbb{N}$ such that $k \geq 2$ or $w_{2\ell}(v, v) > 0$, the following inequality holds:

$$cl_{2\ell+p}(v)^k \leq cl_{2\ell}(v)^{k-1} \cdot cl_{2\ell+pk}(v) \ .$$

For $cl_{2\ell}(v) > 0$ and $cl_{2\ell+p}(v) > 0$, respectively, this is equivalent to

$$\left(\frac{cl_{2\ell+p}(v)}{cl_{2\ell}(v)} \right)^k \leq \frac{cl_{2\ell+pk}(v)}{cl_{2\ell}(v)} \quad and \quad \left(\frac{cl_{2\ell+p}(v)}{cl_{2\ell}(v)} \right)^{k-1} \leq \frac{cl_{2\ell+pk}(v)}{cl_{2\ell+p}(v)} \ .$$

4.4.2 A Special Case for the Average Number of Walks

As a special case of Corollary 4.11 (for $\ell = 0$), we obtain an inequality which compares the average number of walks (per vertex) of lengths p and pk.

Corollary 4.14
For every graph $G = (V, E)$, vertex subset $S \subseteq V$ with $|S| \geq 1$, and $k, p \in \mathbb{N}$, the following inequalities hold:

$$w_p(S, S)^k \leq |S|^{k-1} w_{pk}(S, S) \qquad and \qquad \left(\frac{w_p(S, S)}{|S|}\right)^k \leq \frac{w_{pk}(S, S)}{|S|} \ .$$

For the case $\ell = 0$ and $p = 1$, we obtain $w_1(S, S)^k \leq w_k(S, S) \cdot w_0(S, S)^{k-1}$ where $w_1(S, S)$ is the number of edges in the subgraph induced by S and $w_1(S, S)/w_0(S, S) = w_1(S, S)/|S|$ is the average degree in this subgraph.

If $S = V$, Corollary 4.14 is the following statement for the total number of walks.

Corollary 4.15
For every graph on $n \geq 1$ vertices and $k, p \in \mathbb{N}$, the following inequalities hold:

$$w_p^k \leq n^{k-1} w_{pk} \qquad and \qquad \left(\frac{w_p}{n}\right)^k \leq \frac{w_{pk}}{n} \ .$$

For $\ell = 0$ and $p = 1$, this means $w_1^k \leq w_k \cdot w_0^{k-1}$ which is (by $w_1/w_0 = 2m/n = \bar{d}$) exactly Theorem 4.13.

For the number of closed walks, we do not conclude something new from Corollary 4.14 since if S consists of a single vertex $v \in V$, then we obtain $cl_p(v)^k \leq cl_{pk}(v)$, but this is obvious anyway.

4.4.3 The Density Implication

Graph Density.

Corollary 4.12 implies

$$\frac{w_{2\ell+p}^k}{[n(n-1)^{2\ell+p}]^k} \leq \frac{w_{2\ell}^{k-1} \cdot w_{2\ell+pk}}{[n(n-1)^{2\ell}]^{k-1} \cdot n(n-1)^{2\ell+pk}} \ .$$

Corollary 4.16
For every graph and $k, \ell, p \in \mathbb{N}$, the following inequality holds:

$$\rho_{2\ell+p}^k \leq \rho_{2\ell}^{k-1} \cdot \rho_{2\ell+pk} \ .$$

This extends the known relations (see Kosub [Kos05]) and includes as special cases

$$\rho_p^k \leq \rho_{pk} \quad (\ell = 0) \qquad and \qquad \rho^k \leq \rho_k \quad (\ell = 0, p = 1) \ .$$

Matrix Density.

For nonnegative symmetric matrices, we obtain the following bound for the average row sum:

$$\left(\frac{1}{n}\sum_{i=1}^{n}r_i(A)\right)^k \le \frac{1}{n}\sum_{i=1}^{n}r_i\left(A^k\right)$$

4.4.4 Other Theorems Involving Powers

Results that are similar to the other theorems of this section can be obtained using Rogers' and Hölder's inequality. Provided that we are given a Hermitian matrix, and we derive its (real) eigenvalues λ_i and the corresponding (nonnegative) constants $\hat{c}_{s,i}$ from the spectral decomposition as we did in the previous section, then we can apply Theorem 1.4. For example, if we apply it with $p = \frac{x}{x+y}$ and $q = \frac{y}{x+y}$ to the vector of the numbers $\hat{c}_{s,i}^{x/(x+y)}\lambda_i^a$ and $\hat{c}_{s,i}^{y/(x+y)}\lambda_i^b$ (for $i \in [n]$), we obtain the following calculation:

$$\left(\sum_{i=1}^{n}\left|\hat{c}_{s,i}^{\frac{x}{x+y}}\lambda_i^a\right|\cdot\left|\hat{c}_{s,i}^{\frac{y}{x+y}}\lambda_i^b\right|\right)^{x+y} \le \left(\sum_{i=1}^{n}\left|\hat{c}_{s,i}^{\frac{x}{x+y}}\lambda_i^a\right|^{\frac{x+y}{x}}\right)^{x}\left(\sum_{i=1}^{n}\left|\hat{c}_{s,i}^{\frac{y}{x+y}}\lambda_i^b\right|^{\frac{x+y}{y}}\right)^{y}$$

$$\left(\sum_{i=1}^{n}\hat{c}_{s,i}\lambda_i^{a+b}\right)^{x+y} \le \left(\sum_{i=1}^{n}\hat{c}_{s,i}\left|\lambda_i^{a+b}\right|\right)^{x+y} \le \left(\sum_{i=1}^{n}\hat{c}_{s,i}\left|\lambda_i^a\right|^{\frac{x+y}{x}}\right)^{x}\left(\sum_{i=1}^{n}\hat{c}_{s,i}\left|\lambda_i^b\right|^{\frac{x+y}{y}}\right)^{y}.$$

As indicated in the last line, the absolute values can be left out on the lesser side. On the other side, the absolute values must be handled explicitly if we want to obtain terms that are equal to the quadratic forms we are aiming for.

Positive-Semidefinite Matrices.

Of course, the easiest way to get rid of the absolute value operations is to assume that the eigenvalues are nonnegative, that is, the matrix must be positive-semidefinite. As already mentioned before, it is also possible to define matrix powers for arbitrary real exponents in this case.

Theorem 4.29
For any positive-semidefinite matrix P, the inequality

$$\left[\text{sum}_s\left(P^{a+b}\right)\right]^{x+y} \le \left[\text{sum}_s\left(P^{a\frac{x+y}{x}}\right)\right]^{x}\left[\text{sum}_s\left(P^{b\frac{x+y}{y}}\right)\right]^{y}$$

holds for each complex vector $s \in \mathbb{C}^n$ and real numbers a, b, $x \neq 0$, and $y \neq 0$.

This is a generalization of Theorem 4.23, which is in turn a more general form of Theorem 4.21.

Hermitian Matrices.

For the application to eigenvalues that are allowed to be negative, $a(x + y)/x = a + ay/x$ and $b(x + y)/y = b + bx/y$ have to be even numbers. Further, λ_i^{a+b} has to be defined within the real numbers for all $i \in [n]$. This is true, for instance, if $a + b$ is an integer. (Other possibilities were discussed already in the preceding sections.) For example, if we assume that

$$\frac{a(x + y)}{x} = 2k \qquad \text{and} \qquad \frac{b(x + y)}{y} = 2\ell$$

then we have

$$a + b = \frac{2kx + 2\ell y}{x + y} = \frac{2(kx + \ell y)}{x + y} .$$

As a first special case, we could consider $a = 0$, wich implies $k = 0$. We obtain

$$\text{sum}_s \left(A^{\frac{2\ell y}{x+y}} \right)^{x+y} \leq \text{sum}_s \left(A^0 \right)^x \cdot \text{sum}_s \left(A^{2\ell} \right)^y .$$

If we apply this to the number of walks in undirected graphs, this would translate to

$$w^{x+y}_{\frac{2\ell y}{x+y}} \leq w_0^x \cdot w_{2\ell}^y .$$

This is a generalization of Theorem 4.15 (Godsil, see [ES82]) since it also includes cases with odd exponents on the lesser side. An example would be $w_2^3 \leq w_0^2 \cdot w_6$ or $(w_2/n)^3 \leq w_6/n$. On the other hand, this result can also be proven by application of Theorem 4.20 (Mulholland and Smith [MS59]) to $B = A^2$, $v = \mathbf{1}_n$, and $k = 3$.

For the special case $k = 1$, we obtain

$$\text{sum}_s \left(A^{\frac{2x+2\ell y}{x+y}} \right)^{x+y} \leq \text{sum}_s \left(A^2 \right)^x \cdot \text{sum}_s \left(A^{2\ell} \right)^y$$

and

$$w^{x+y}_{\frac{2x+2\ell y}{x+y}} \leq w_2^x \cdot w_{2\ell}^y .$$

Closed Walks.

Using an equivalent approach for the trace of the matrix powers (which corresponds to the number of closed walks in undirected graphs), we obtain

$$\left(\sum_{i=1}^n \lambda_i^{a+b} \right)^{x+y} \leq \left(\sum_{i=1}^n |\lambda_i^{a+b}| \right)^{x+y} \leq \left(\sum_{i=1}^n |\lambda_i^a|^{\frac{x+y}{x}} \right)^x \left(\sum_{i=1}^n |\lambda_i^b|^{\frac{x+y}{y}} \right)^y .$$

4.4.5 The Method of Expectation of Random Variables

Similar results as in Subsection 4.2.2 can be obtained using Lyapunov's inequality (see Karr [Kar93]):

$$\left[\mathbb{E}\left(|X|^a \right) \right]^{1/a} \leq \left[\mathbb{E}\left(|X|^b \right) \right]^{1/b} \qquad \text{for } 1 \leq a \leq b .$$

For even number $b = 2b'$, $b' \in \mathbb{N}$, we obtain

$$\left[\mathbb{E}(X^a)\right]^{1/a} \leq \left[\mathbb{E}(|X|^a)\right]^{1/a} \leq \left[\mathbb{E}(X^{2b'})\right]^{1/(2b')} \quad \text{for } 1 \leq a \leq b \; .$$

or

$$\left(\frac{\mathrm{sum}_s\left(A^{2k+a}\right)}{\mathrm{sum}_s\left(A^{2k}\right)}\right)^{1/a} \leq \left(\frac{\mathrm{sum}_s\left(A^{2k+2b'}\right)}{\mathrm{sum}_s\left(A^{2k}\right)}\right)^{1/2b'} \; .$$

This implies a statement that is similar to the theorem of Mulholland and Smith [MS59] (see Theorem 4.20).

Theorem 4.30
For all Hermitian matrices A, integers $a, b, k \in \mathbb{N}$ with $1 \leq a \leq 2b$, and scaling vectors $s \in \mathbb{C}^n$, the following inequality holds:

$$\left[\mathrm{sum}_s\left(A^{2k+a}\right)\right]^{2b} \leq \left[\mathrm{sum}_s\left(A^{2k}\right)\right]^{2b-a} \cdot \left[\mathrm{sum}_s\left(A^{2k+2b}\right)\right]^{a} \; .$$

Corollary 4.17
For all Hermitian matrices A, integers $a, b, k \in \mathbb{N}$ with $1 \leq a \leq 2b$, and subsets $S \subseteq [n]$, the following inequality holds:

$$\left[\mathrm{sum}\left(A^{2k+a}[S]\right)\right]^{2b} \leq \left[\mathrm{sum}\left(A^{2k}[S]\right)\right]^{2b-a} \cdot \left[\mathrm{sum}\left(A^{2k+2b}[S]\right)\right]^{a} \; .$$

In particular, if $S = [n]$, we have

$$\left[\mathrm{sum}\left(A^{2k+a}\right)\right]^{2b} \leq \left[\mathrm{sum}\left(A^{2k}\right)\right]^{2b-a} \cdot \left[\mathrm{sum}\left(A^{2k+2b}\right)\right]^{a} \; ,$$

and if $S = \{i\}$ with $v \in [n]$, we have

$$\left(a_{i,i}^{[2k+a]}\right)^{2b} \leq \left(a_{i,i}^{[2k]}\right)^{2b-a} \cdot \left(a_{i,i}^{[2k+2b]}\right)^{a} \; .$$

Corollary 4.18
For every undirected graph $G = (V, E)$, integers $a, b, k \in \mathbb{N}$ with $1 \leq a \leq 2b$, and subsets $S \subseteq V$, the following inequality holds:

$$[w_{2k+a}(S, S)]^{2b} \leq [w_{2k}(S, S)]^{2b-a} \cdot [w_{2k+2b}(S, S)]^{a} \; .$$

In particular, if $S = V$, we have

$$w_{2k+a}^{2b} \leq w_{2k}^{2b-a} \cdot w_{2k+2b}^{a} \; ,$$

and if $S = \{v\}$ with $v \in V$, we have

$$[cl_{2k+a}(v)]^{2b} \leq [cl_{2k}(v)]^{2b-a} \cdot [cl_{2k+2b}(v)]^{a} \; .$$

The special case where $S = V$ and $k = 0$ corresponds to Godsil's Theorem 4.15.

For positive-semidefinite matrices, we can ignore the absolute values and we obtain

$$\left[\mathbb{E}(X^a)\right]^{1/a} \leq \left[\mathbb{E}(X^b)\right]^{1/b} \quad \text{for } 1 \leq a \leq b \; .$$

Theorem 4.31

For all positive-semidefinite matrices P, real numbers $a, b, k \in \mathbb{R}$ with $1 \le a \le b$, and scaling vectors $s \in \mathbb{C}^n$, the following inequality holds:

$$\left[\mathrm{sum}_s \left(P^{k+a} \right) \right]^b \le \left[\mathrm{sum}_s \left(P^k \right) \right]^{b-a} \cdot \left[\mathrm{sum}_s \left(P^{k+b} \right) \right]^a .$$

4.5 The Number of Walks and Degree Powers

4.5.1 Related Work

Relevant research regarding the powers of vertex degrees in graphs can be split into two main branches. One line of research tried to find upper and lower bounds for the sum of degree powers in graphs with fixed number of vertices and number of edges (or with other restrictions like a fixed degree sequence). The other line of research focused on characterizing the extremal graphs. The development started with investigations of the sum of squared degrees.

Ahlswede and Katona [AK78] investigated the graphs with the maximum number of adjacent pairs of edges, and thus with the maximum number of 2-step walks, for given numbers of vertices and edges. Note that their function is based on pairs of *different* edges, but this maximizes the same as w_2 since it always ignores the m pairs of edges consisting of twice the same edge (which corresponds to the $2m$ walks of length 2 that traverse the same edge in opposing directions).

For $p \in \mathbb{N} \setminus \{0\}$, Székely, Clark and Entringer [SCE92] proved that

$$\sum_{v \in V} d(v)^p \le \left(\sum_{v \in V} d(v)^{1/p} \right)^p .$$

For degree squares, de Caen [dCae98] showed the inequality

$$\sum_{v \in V} d(v)^2 \le m \left(\frac{2m}{n-1} + n - 2 \right) \qquad \text{(for } n \ge 2\text{)} ,$$

which is incomparable to the inequality of Székely et al. It implies for all trees the bound $\sum_{v \in V} d(v)^2 \le n(n-1)$. De Caen's inequality was applied by Li and Pan [LP01] to find an upper bound for the largest Laplacian eigenvalue of a graph.

Nikiforov [Nik07d] showed the following inequality:

$$\sum_{v \in V} d(v)^2 \le \begin{cases} (2m)^{3/2} & \text{for } m \ge n^2/4 \\ (n^2 - 2m)^{3/2} + 4mn - n^3 & \text{for } m < n^2/4 \end{cases} .$$

Other upper and lower bounds for the sum of squared vertex degrees, as well as graphs that achieve the bounds and graphs that maximize or minimize the sum were discussed in the articles by Peled, Petreschi and Sterbini [PPS99], Gutman [Gut03], Das [Das03; Das04], Ábrego, Fernández-Merchant, Neubauer and Watkins [ÁFNW09], Wagner and Wang [WW09] and Gutman [Gut14a].

The k-th moment of the degree sequence

$$\mu_k = \frac{1}{n} \sum_{v \in V} d(v)^k$$

was discussed in the papers by Füredi and Kündgen [FK06] and Cioabă [Cio06]. Cioabă [Cio06] used the equality

$$\sum_{v \in V} d(v)^{k+1} = \sum_{v \in V} d(v) \cdot m_k(v)$$

(where $m_k(v) = \sum_{\{v,w\} \in E} d(w)^k / d(v)$ is the average of the k-th powers of the degrees of the neighbors of v) to deduce the following inequality using Chebyshev's inequality (see Theorem 1.6):

$$\sum_{v \in V} d(v)^{k+1} \geq \frac{2m}{n} \sum_{v \in V} d(v)^k \ .$$

Let us remark here that it is easy to obtain the following inequality using the same argument:

$$\frac{1}{n} \sum_{v \in V} d(v)^2 \cdot d(v)^k \ \geq \ \frac{1}{n} \sum_{v \in V} d(v)^2 \cdot \frac{1}{n} \sum_{v \in V} d(v)^k$$

$$\sum_{v \in V} d(v)^{k+2} \ \geq \ \frac{w_2}{n} \cdot \sum_{v \in V} d(v)^k \ .$$

Both results can be generalized to row sums of symmetric matrices.

The following inequality for undirected graphs was conjectured by Noy and proven by Fiol and Garriga [FG09].

Theorem 4.32 (Fiol and Garriga)
For every undirected graph, the number w_k of walks of length k does not exceed the sum of the k-th powers of the vertex degrees, i.e.,

$$w_k \leq \sum_{v \in V} d(v)^k \ .$$

The closely related graph homomorphism numbers for paths and stars are discussed at the very end of this work. Several applications for the sum of vertex degree powers are discussed in the article of Cao, Dehmer and Shi [CDS14]. In particular, a corresponding graph entropy measure was considered.

Actually, Theorem 4.32 is a special case (for adjacency matrices) of a much older theorem for powers of nonnegative symmetric matrices and their row or column sums which was conjectured by London [Lon66b] and proven by Hoffman [Hof67].

Theorem 4.33 (London; Hoffman)
For every symmetric nonnegative matrix and $p \in \mathbb{N}$ holds:

$$\mathrm{sum}\,(A^p) \leq \sum_{i=1}^{n} r_i^p$$

Another proof of this theorem has been published by Sidorenko [Sid85b; Sid85a]. He also showed that for $k > 1$ equality is achieved if and only if A is decomposable into a direct sum of matrices which are proportional to doubly stochastic matrices.

4.5.2 Refined Inequalities for Powers of Degrees or Row Sums

Now, we refine the inequality by Fiol and Garriga as follows.

Theorem 4.34
For all undirected graphs and $p, q \in \mathbb{N}$ with $p \geq 1$, we have

$$\sum_{v \in V} d(v)^q w_p(v) \leq \sum_{v \in V} d(v)^{q+1} w_{p-1}(v) \ .$$

In particular, this implies the following.

Corollary 4.19
For all undirected graphs and $p, q \in \mathbb{N}$, we have

$$w_{p+q} \leq \sum_{v \in V} d(v)^q w_p(v) \leq \sum_{v \in V} d(v)^{p+q} \ .$$

More generally, we will refine the inequality of London and Hoffman as follows.

Theorem 4.35
For every nonnegative symmetric matrix A and $p, q \in \mathbb{N}$ with $p \geq 1$, we have

$$\sum_{i=1}^{n} r_i(A)^q \cdot r_i(A^p) \leq \sum_{i=1}^{n} r_i(A)^{q+1} \cdot r_i(A^{p-1}) \ .$$

This corresponds to the short form $\sum_{i=1}^{n} r_i^q r_i^{[p]} \leq \sum_{i=1}^{n} r_i^{q+1} r_i^{[p-1]}$.

Corollary 4.20
For every nonnegative symmetric matrix A and $p, q \in \mathbb{N}$, we have

$$\mathrm{sum}\left(A^{p+q}\right) \leq \sum_{i=1}^{n} r_i(A)^q \cdot r_i(A^p) \leq \sum_{i=1}^{n} r_i(A)^{p+q} \ .$$

This corresponds to the short form $\mathrm{sum}\left(A^{p+q}\right) \leq \sum_{i=1}^{n} r_i^q r_i^{[p]} \leq \sum_{i=1}^{n} r_i^{p+q}$. We will prove these inequalities in an even more general setting for directed graphs and nonsymmetric matrices, see Theorems 7.7 and 7.8 in Chapter 7.

4.5.3 Other Degree-Related Inequalities

Now we show some inequalities that will turn out to be very useful later (for proving results on restricted graph classes).

Lemma 4.1
Every graph $G = (V, E)$ satisfies the inequality

$$w_{k+1}^2 \leq m \sum_{\{x,y\} \in E} [w_k(x) + w_k(y)]^2$$

Proof   By application of Lemma 1.4, we have

$$\left(\sum_{\{x,y\} \in E} (w_k(x) + w_k(y)) \right)^2 \leq m \sum_{\{x,y\} \in E} [w_k(x) + w_k(y)]^2 ,$$

$$w_{k+1}^2 \leq m \sum_{\{x,y\} \in E} [w_k(x) + w_k(y)]^2 .$$

The proof is complete by observing $w_{k+1}(G) = \sum_{v \in V} d_v w_k(v) = \sum_{\{x,y\} \in E} (w_k(x) + w_k(y))$ (for loop-free graphs).

Corollary 4.21
Every graph $G = (V, E)$ satisfies the inequality

$$w_2(G)^2 \leq m \sum_{\{x,y\} \in E} (d_x + d_y)^2 .$$

Now we show the main results of this subsection.

Theorem 4.36
Every graph $G = (V, E)$ satisfies the inequality

$$2w_1 w_2 \leq w_0 \sum_{\{x,y\} \in E} (d_x + d_y)^2 = w_0 \left(w_3 + \sum_{v \in V} d_v^3 \right) .$$

Proof   For the proof, we use the equivalent form

$$2\frac{w_1}{w_0} \leq \frac{1}{w_2} \sum_{\{x,y\} \in E} (d_x + d_y)^2 .$$

We prove the inequality by separating both sides using the term w_2/m. First, we have $2w_1/w_0 \leq w_2/m$. This is obviously true since $w_1 = 2m$ and $w_1^2 \leq w_0 w_2$ by Theorem 4.1. It remains to show that

$$\frac{w_2}{m} \leq \frac{1}{w_2} \sum_{\{x,y\} \in E} (d_x + d_y)^2 ,$$

which is true by Corollary 4.21. The equality part results from $\sum_{\{x,y\} \in E} (d_x + d_y)^2 = \sum_{\{x,y\} \in E} d_x^2 + d_y^2 + 2d_x d_y = 2(\sum_{\{x,y\} \in E} d_x d_y) + \sum_{v \in V} d_v^3 = w_3 + \sum_{v \in V} d_v^3$.

It is interesting to see that the inequality

$$w_1 w_2 \leq w_0 \frac{w_3 + \sum_{v \in V} d_v^3}{2}$$

holds true although there are graphs with $w_1 w_2 \not\leq w_0 w_3$ and there are (other) graphs satisfying $w_3 = \sum_{v \in V} d_v^3$. Another interesting equivalent form uses arithmetic means:

$$\frac{1}{n} \sum_{x \in V} d_x^2 \leq \frac{1}{m} \sum_{\{x,y\} \in E} \left(\frac{d_x + d_y}{2} \right)^2 .$$

While arithmetic means appear on the left side (per vertex) and on the right side (per edge), there is also the squared mean of d_x and d_y on the right handside. Replacing this arithmetic mean $(d_x + d_y)/2$ by the smaller geometric mean $\sqrt{d_x d_y}$ would lead to the inequality $w_1 w_2 \leq w_0 w_3$, which is not valid for general graphs (as discussed earlier).

4.6 Invalid Inequalities

4.6.1 Known Results

Marcus and Newman [MN62, p. 634] made the conjecture that for all nonnegative symmetric matrices A, we would have

$$\mathrm{sum}(A) \cdot \mathrm{sum}(A^2) \leq n \cdot \mathrm{sum}(A^3) .$$

It was then shown by London [Lon66b] and Kankaanpää and Merikoski [KM84] that in the case of even exponent m the more general conjecture

$$\mathrm{sum}(A) \cdot \mathrm{sum}(A^m) \leq n \cdot \mathrm{sum}(A^{m+1})$$

holds true only for matrices of dimension $n \leq 3$. (For odd exponent m, the conjecture is true in general, see Theorem 4.4.) The inequality also holds for $n \geq 4$ if the row sums of A and A^2 are similarly ordered. This follows from Chebyshev's inequality. The inequality is reversed if the row sums of A and A^2 are conversely ordered. For $n \geq 4$ and an even number $m \geq 2$, London constructed a family of matrices showing $\mathrm{sum}(A) \cdot \mathrm{sum}(A^m) \not\leq n \cdot \mathrm{sum}(A^{m+1})$, which disproves the conjecture of Marcus and Newman.

Theorem 4.37 (London)
For $k \in \mathbb{N} \setminus \{0\}$ and $n \geq 4$, there is a nonnegative symmetric $n \times n$ matrix A that shows

$$\mathrm{sum}(A) \cdot \mathrm{sum}(A^{2k}) \not\leq n \cdot \mathrm{sum}(A^{2k+1}) .$$

Note that these counterexamples are nonnegative symmetric matrices, not adjacency matrices.

Lagarias, Mazo, Shepp and McKay [LMSM84] disproved the inequality $w_r \cdot w_s \leq n \cdot w_{r+s}$ for all pairs (r, s) where the sum $r + s$ takes an odd value.

Theorem 4.38 (Lagarias et al.)
For all $r, s \in \mathbb{N}$ where $r + s$ is odd, there is a graph that shows

$$w_r \cdot w_s \nleq n \cdot w_{r+s} \ .$$

Although the authors do not cite any references, we note that this strengthens London's result in two ways. The theorem shows that there are counterexamples even in the very restricted case where the matrices are adjacency matrices of graphs (but the matrices of the counterexamples are larger than before), and they apply to more general exponents (walk lengths). The counterexamples were disconnected graphs consisting of a complete graph K_{m+1} and a star S_{m^2+t+1} for $t \geq 1$ and m sufficiently large. They noted that connected counterexamples could be constructed by adding an edge between a vertex of the complete subgraph and a leaf of the star. As counterexamples for the most basic inequality $w_1 w_2 \leq n w_3$, they proposed the disconnected graph obtained from the disjoint union of K_3 and S_6. As a connected counterexample, they proposed to connect the complete graph K_9 via a new edge to a leaf of the star S_{89}.[5]

4.6.2   Extended Results

We will reuse the method of Lagarias et al. and try to construct counterexamples for other possible inequalities. In particular, we consider the conceivable sandwich inequality

$$w_{a+c} w_{a+2b+c+1} \overset{?}{\leq} w_a w_{a+2b+2c+1} \ .$$

As can be seen, each side consists of a product involving one even and one odd walk length. Again, the graphs consist of a complete graph K_{m+1} and a star S_{m^2+t+1} for $t \geq 1$. For a complete graph on n vertices, we have $w_i = n(n-1)^i$. For the number of walks of the star part, we have to distinguish between even and odd walk lengths. For a star on n vertices, we have $w_{2i} = n(n-1)^i$ and $w_{2i+1} = 2(n-1)^{i+1}$, see Section 5.3. Now we need to distinguish whether a and c take odd or even values.

Note that in any case we have $w_{a+c}(G_i) w_{a+2b+c+1}(G_i) = w_a(G_i) w_{a+2b+2c+1}(G_i)$ for each of the two subgraphs $G_1 = K_{m+1}$ and $G_2 = S_{m^2+t+1}$. Thus, only the mixed terms need to be considered for the difference.

$$
\begin{aligned}
& w_{a+c} w_{a+2b+c+1} - w_a w_{a+2b+2c+1} \\
= \ & w_{a+c}(K_{m+1}) w_{a+2b+c+1}(S_{m^2+t+1}) + w_{a+2b+c+1}(K_{m+1}) w_{a+c}(S_{m^2+t+1}) \\
& - w_a(K_{m+1}) w_{a+2b+2c+1}(S_{m^2+t+1}) - w_{a+2b+2c+1}(K_{m+1}) w_a(S_{m^2+t+1})
\end{aligned}
$$

---

[5]There is a mistake in [LMSM84] since setting $m = 8$, $t = 7$ leads to K_9 connected to S_{72}, not S_{89}. Later, we will see a smaller connected counterexample (19 vertices) than this one.

Case 1:

a even ($a + 2b + 2c + 1$ odd), c odd ($a + c$ odd , $a + 2b + c + 1$ even)

	K_{m+1}	S_{m^2+t+1}
w_{a+c}	$(m+1)m^{a+c}$	$2(m^2+t)^{(a+c+1)/2}$
$w_{a+2b+c+1}$	$(m+1)m^{a+2b+c+1}$	$(m^2+t+1)(m^2+t)^{(a+2b+c+1)/2}$
w_a	$(m+1)m^a$	$(m^2+t+1)(m^2+t)^{a/2}$
$w_{a+2b+2c+1}$	$(m+1)m^{a+2b+2c+1}$	$2(m^2+t)^{(a+2b+2c+2)/2}$

$$
\begin{aligned}
w_{a+c}w_{a+2b+c+1} - w_a w_{a+2b+2c+1} =\ & (m+1)m^{a+c}(m^2+t+1)(m^2+t)^{(a+2b+c+1)/2} \\
& +(m+1)m^{a+2b+c+1} \cdot 2(m^2+t)^{(a+c+1)/2} \\
& -(m+1)m^a \cdot 2(m^2+t)^{(a+2b+2c+2)/2} \\
& -(m+1)m^{a+2b+2c+1}(m^2+t+1)(m^2+t)^{a/2}
\end{aligned}
$$

$$
\begin{aligned}
& \frac{w_{a+c}w_{a+2b+c+1} - w_a w_{a+2b+2c+1}}{(m+1)m^a(m^2+t)^{a/2}} \\
=\ & m^c(m^2+t+1)\left[(m^2+t)^{(2b+c+1)/2} - m^{2b+c+1}\right] \\
& +2(m^2+t)^{(c+1)/2}\left[m^{2b+c+1} - (m^2+t)^{(2b+c+1)/2}\right] \\
=\ & \left[m^c(m^2+t+1) - 2(m^2+t)^{(c+1)/2}\right]\left[(m^2+t)^{(2b+c+1)/2} - m^{2b+c+1}\right] \\
=\ & \left[m^{c+2} + m^c(t+1) - 2(m^2+t)^{(c+1)/2}\right]\left[(m^2+t)^{(2b+c+1)/2} - m^{2b+c+1}\right]
\end{aligned}
$$

By the binomial theorem, we have $(m^2 + t)^{(2b+c+1)/2} - m^{2b+c+1} > 0$ and $2(m^2 + t)^{(c+1)/2} \in \mathcal{O}(m^{c+1})$. Therefore, the difference must be strictly positive for fixed $t \geq 1$ and sufficiently large m. The inequality at the beginning of this subsection does not hold.

Case 2:

a even ($a + 2b + 2c + 1$ odd), c even ($a + c$ even , $a + 2b + c + 1$ odd)

	K_{m+1}	S_{m^2+t+1}
w_{a+c}	$(m+1)m^{a+c}$	$(m^2+t+1)(m^2+t)^{(a+c)/2}$
$w_{a+2b+c+1}$	$(m+1)m^{a+2b+c+1}$	$2(m^2+t)^{(a+2b+c+2)/2}$
w_a	$(m+1)m^a$	$(m^2+t+1)(m^2+t)^{a/2}$
$w_{a+2b+2c+1}$	$(m+1)m^{a+2b+2c+1}$	$2(m^2+t)^{(a+2b+2c+2)/2}$

$$
\begin{aligned}
w_{a+c}w_{a+2b+c+1} - w_a w_{a+2b+2c+1} =\ & (m+1)m^{a+c}2(m^2+t)^{(a+2b+c+2)/2} \\
& +(m+1)m^{a+2b+c+1}(m^2+t+1)(m^2+t)^{(a+c)/2} \\
& -(m+1)m^a 2(m^2+t)^{(a+2b+2c+2)/2} \\
& -(m+1)m^{a+2b+2c+1}(m^2+t+1)(m^2+t)^{a/2}
\end{aligned}
$$

$$\frac{w_{a+c}w_{a+2b+c+1} - w_a w_{a+2b+2c+1}}{(m+1)m^a(m^2+t)^{a/2}}$$

$$= m^{2b+c+1}(m^2+t+1)\left[(m^2+t)^{c/2} - m^c\right]$$

$$+ 2(m^2+t)^{(2b+c+2)/2}\left[m^c - (m^2+t)^{c/2}\right]$$

$$= \left[m^{2b+c+1}(m^2+t+1) - 2(m^2+t)^{(2b+c+2)/2}\right]\left[(m^2+t)^{c/2} - m^c\right]$$

$$= \left[m^{2b+c+3} + m^{2b+c+1}(t+1) - 2(m^2+t)^{(2b+c+2)/2}\right]\left[(m^2+t)^{c/2} - m^c\right]$$

By the binomial theorem, we have $(m^2+t)^{c/2} - m^c > 0$ and $2(m^2+t)^{(2b+c+2)/2} \in \mathcal{O}(m^{2b+c+2})$. Therefore, the difference must be strictly positive for fixed $t \geq 1$ and sufficiently large m. The inequality at the beginning of this subsection does not hold.

Case 3:

a odd ($a + 2b + 2c + 1$ even), c even ($a + c$ odd, $a + 2b + c + 1$ even)

	K_{m+1}	S_{m^2+t+1}
w_{a+c}	$(m+1)m^{a+c}$	$2(m^2+t)^{(a+c+1)/2}$
$w_{a+2b+c+1}$	$(m+1)m^{a+2b+c+1}$	$(m^2+t+1)(m^2+t)^{(a+2b+c+1)/2}$
w_a	$(m+1)m^a$	$2(m^2+t)^{(a+1)/2}$
$w_{a+2b+2c+1}$	$(m+1)m^{a+2b+2c+1}$	$(m^2+t+1)(m^2+t)^{(a+2b+2c+1)/2}$

$$
\begin{aligned}
w_{a+c}w_{a+2b+c+1} - w_a w_{a+2b+2c+1} &= (m+1)m^{a+c}(m^2+t+1)(m^2+t)^{(a+2b+c+1)/2} \\
&+ (m+1)m^{a+2b+c+1} \cdot 2(m^2+t)^{(a+c+1)/2} \\
&- (m+1)m^a(m^2+t+1)(m^2+t)^{(a+2b+2c+1)/2} \\
&- (m+1)m^{a+2b+2c+1} \cdot 2(m^2+t)^{(a+1)/2}
\end{aligned}
$$

$$\frac{w_{a+c}w_{a+2b+c+1} - w_a w_{a+2b+2c+1}}{(m+1)m^a(m^2+t)^{a/2}}$$

$$= (m^2+t)^{(2b+c+1)/2}(m^2+t+1)\left[m^c - (m^2+t)^{c/2}\right]$$

$$+ m^{2b+c+1} \cdot 2(m^2+t)^{1/2}\left[(m^2+t)^{c/2} - m^c\right]$$

$$= \left[(m^2+t)^{c/2} - m^c\right]\left[m^{2b+c+1} \cdot 2(m^2+t)^{1/2} - (m^2+t)^{(2b+c+1)/2}(m^2+t+1)\right]$$

The last term is negative. We do not obtain a counterexample in this case.

Case 4:

a odd ($a + 2b + 2c + 1$ even), c odd ($a + c$ even, $a + 2b + c + 1$ odd)

	K_{m+1}	S_{m^2+t+1}
w_{a+c}	$(m+1)m^{a+c}$	$(m^2 + t + 1)(m^2 + t)^{(a+c)/2}$
$w_{a+2b+c+1}$	$(m+1)m^{a+2b+c+1}$	$2(m^2 + t)^{(a+2b+c+2)/2}$
w_a	$(m+1)m^a$	$2(m^2 + t)^{(a+1)/2}$
$w_{a+2b+2c+1}$	$(m+1)m^{a+2b+2c+1}$	$(m^2 + t + 1)(m^2 + t)^{(a+2b+2c+1)/2}$

$$
\begin{aligned}
w_{a+c}w_{a+2b+c+1} - w_a w_{a+2b+2c+1} = \; & (m+1)m^{a+c} \cdot 2(m^2 + t)^{(a+2b+c+2)/2} \\
& + (m+1)m^{a+2b+c+1}(m^2 + t + 1)(m^2 + t)^{(a+c)/2} \\
& - (m+1)m^a(m^2 + t + 1)(m^2 + t)^{(a+2b+2c+1)/2} \\
& - (m+1)m^{a+2b+2c+1} \cdot 2(m^2 + t)^{(a+1)/2}
\end{aligned}
$$

$$
\begin{aligned}
& \frac{w_{a+c}w_{a+2b+c+1} - w_a w_{a+2b+2c+1}}{m^a(m+1)(m^2+t)^{a/2}} \\
= \; & (m^2 + t)^{c/2}(m^2 + t + 1)\left[m^{2b+c+1} - (m^2 + t)^{(2b+c+1)/2} \right] \\
& + m^c \cdot 2(m^2 + t)^{1/2}\left[(m^2 + t)^{(2b+c+1)/2} - m^{2b+c+1} \right] \\
= \; & \left[(m^2 + t)^{(2b+c+1)/2} - m^{2b+c+1} \right]\left[m^c \cdot 2(m^2 + t)^{1/2} - (m^2 + t)^{c/2}(m^2 + t + 1) \right]
\end{aligned}
$$

The last term is negative. We do not obtain a counterexample in this case.

In summary, we can state the following for the case of odd-times-even lengths on both sides when a is even ($a = 2a'$).

Theorem 4.39
If $a', b, c \in \mathbb{N}$ then there are graphs with

$$
w_{2a'+c}w_{2a'+2b+c+1} \nleq w_{2a'}w_{2a'+2b+2c+1} \; .
$$

This generalizes Theorem 4.38. We do not know whether the inequality is valid in the other case.

Now it would be interesting to prove or disprove those inequalities for restricted graph classes, such as the class of all bipartite graphs, the class of all cycle-free graphs (forests), and the class of trees. This will be discussed in Chapter 5.

4.7 Relaxed Inequalities Using Geometric Means

As we have seen, there are graphs with $w_1 w_k \nleq w_0 w_{k+1}$ for even $k \in \mathbb{N} \setminus \{0\}$. Recall that this corresponds to the inequality $\bar{d} \cdot w_k \nleq w_{k+1}$, which uses the average degree of the vertices. The question that arises now is whether the inequality would be valid if the arithmetic mean $\bar{d} = \frac{1}{n}\sum_{v \in V} d(v)$ is replaced by another mean. For instance, for the geometric mean of the vertex degrees, this would correspond to

$$w_{k+1} \overset{?}{\geq} \sqrt[n]{\prod_{v \in V} d(v)} \cdot w_k \qquad \text{or} \qquad \left(\frac{w_{k+1}}{w_k}\right)^n \overset{?}{\geq} \prod_{v \in V} d(v) \ .$$

For the harmonic mean of the vertex degrees, this would be

$$w_{k+1} \overset{?}{\geq} \frac{n}{\sum_{i=1}^{n} \frac{1}{d_i}} \cdot w_k \ .$$

At the moment, this problem must be left open, but we provide a weaker result in the next subsection.

We have seen that there are graphs where the product of the average number of 1-step walks and the average number of 2-step walks per vertex is not a valid lower bound for the average number of 3-step walks:

$$\frac{w_1}{n} \cdot \frac{w_2}{n} \not\leq \frac{w_3}{n} \qquad \text{since} \qquad w_1 \cdot w_2 \not\leq n \cdot w_3$$

However, replacing the arithmetic means

$$\underset{v \in V}{\mathfrak{A}} \ w_1(v) = \frac{1}{n}\sum_{v \in V} w_1(v) = \frac{w_1}{n} \qquad \text{and} \qquad \underset{v \in V}{\mathfrak{A}} \ w_2(v) = \frac{1}{n}\sum_{v \in V} w_2(v) = \frac{w_2}{n}$$

by the corresponding geometric means

$$\underset{v \in V}{\mathfrak{G}} \ w_1(v) = \sqrt[n]{\prod_{v \in V} w_1(v)} \qquad \text{and} \qquad \underset{v \in V}{\mathfrak{G}} \ w_2(v) = \sqrt[n]{\prod_{v \in V} w_2(v)}$$

yields a valid inequality. We state it in the more general form for row sums of matrices. Recall that $r_i^{[k]} = r_i\left(A^k\right)$.

Theorem 4.40
For every symmetric matrix with nonnegative row sums, we have

$$\left(\underset{i \in [n]}{\mathfrak{G}} \ r_i^{[k]}\right)\left(\underset{i \in [n]}{\mathfrak{G}} \ r_i^{[\ell]}\right) \leq \underset{i \in [n]}{\mathfrak{A}} \ r_i^{[k+\ell]} = \frac{\operatorname{sum}\left(A^{k+\ell}\right)}{n} \ .$$

Note that for symmetric matrices, we have $r_i = c_i$ and $r_i^{[k]} = c_i^{[k]}$. A generalized result can be obtained for real matrices with nonnegative row and column sums, see Theorem 6.2. There, we give a simple proof using the decomposition lemma and the AM-GM inequality.

Corollary 4.22
For every undirected graph, we have

$$\left(\underset{v \in V}{\mathfrak{G}} \ w_k(v)\right)\left(\underset{v \in V}{\mathfrak{G}} \ w_\ell(v)\right) \leq \underset{v \in V}{\mathfrak{A}} \ w_{k+\ell}(v) = \frac{w_{k+\ell}}{n} \ .$$

A corresponding inequality for directed graphs can be found in Corollary 6.4.

4.7.1 Remarks on Another Relaxed Inequality

There is also another valid relaxation of the generally invalid inequality $w_1 w_2 \not\leq w_0 w_3$. This relaxation considers the following multiplicative variants of the Zagreb indices that were proposed in general by Todeschini and Consonni [TC10] and in this particular variant by Gutman [Gut11].

$$\Pi_1 = \prod_{v \in V} d(v)^2 \qquad \text{and} \qquad \Pi_2 = \prod_{\{u,v\} \in E} d(u) \cdot d(v) .$$

Réti and Gutman [RG12] remarked that Π_1 is just the square of the product of all vertex degrees (also called the Narumi-Katayama index). They showed that the bounds

$$\Pi_1 \leq \left(\frac{2m}{n} \right)^{2n} \qquad \text{and} \qquad \Pi_1 \leq \left(\frac{M_1}{n} \right)^{n}$$

hold for all connected graphs. Actually, both bounds are simple consequences of the inequality of arithmetic and geometric means and they are also valid for disconnected graphs. Réti and Gutman also showed that we have

$$\Pi_2 \geq \left(\frac{2m}{n} \right)^{2m}$$

for all connected graphs.

Notice that the inequality $w_1 w_2 \not\leq w_0 w_3$ can be considered in the form

$$\frac{1}{n} \sum_{v \in V} d(v)^2 \not\leq \frac{1}{m} \sum_{\{u,v\} \in E} d(u) \cdot d(v)$$

where both sides correspond to certain arithmetic means. Now it is a natural question whether the inequality would be valid in general if the arithmetic means are replaced by geometric means. This is in fact true. The corresponding inequality $\sqrt[n]{\Pi_1} \leq \sqrt[m]{\Pi_2}$ has been proven by Eliasi and Vukičević [EV13].

Theorem 4.41 (Eliasi and Vukičević)
For every undirected graph with n vertices and m edges, we have

$$\sqrt[n]{\prod_{v \in V} d(v)^2} \leq \sqrt[m]{\prod_{\{x,y\} \in E} d(x) \cdot d(y)} .$$

Firstly, it should be noticed that this inequality also follows for connected graphs from the two inequalities $\Pi_1 \leq (2m/n)^{2n}$ and $\Pi_2 \geq (2m/n)^{2m}$ mentioned above. On the other hand, we remark that there is a much more compact way to show this. We observe that the left hand side is equal to $\left(\prod_{v \in V} d(v) \right)^{\frac{2}{n}}$ while the right hand side is equal to $\sqrt[m]{\prod_{v \in V} d(v)^{d(v)}}$. That means, the result is equivalent to the following statement.

Corollary 4.23
For every undirected graph $G = (V, E)$ with average degree $\bar{d}$, we have

$$\left(\prod_{v \in V} d(v) \right)^{\bar{d}} \leq \prod_{v \in V} d(v)^{d(v)} .$$

We show the following more general form with a really short proof.

Theorem 4.42
For all nonnegative vectors $a \in \mathbb{R}^n_{\geq 0}$ and $\bar{a} := \frac{1}{n} \sum_{i \in [n]} a_i$, we have

$$\left(\prod_{i=1}^{n} a_i \right)^{\bar{a}} \leq \prod_{i=1}^{n} a_i^{a_i} .$$

Proof   If we ignore the trivial case of zeros, the inequality is equivalent to

$$\bar{a} \ln \prod_{i=1}^{n} a_i \ \leq \ \ln \prod_{i=1}^{n} a_i^{a_i}$$

$$\left(\frac{1}{n} \sum_{i=1}^{n} a_i \right) \sum_{i=1}^{n} \ln a_i \ \leq \ \sum_{i=1}^{n} a_i \ln a_i .$$

Now the result follows directly from Chebyshev's inequality (see Theorem 1.6).

Again, Theorem 4.42 can be regarded as a statement about (weighted) geometric means if it is written as

$$\left(\prod_{i=1}^{n} a_i \right)^{1/n} \leq \left(\prod_{i=1}^{n} a_i^{a_i} \right)^{1/\sum_{i=1}^{n} a_i} .$$

This form (or the form of weighted arithmetic means that results from taking the logarithm) yields an alternative proof of the theorem.

Let us also remark here that we also have

$$\prod_{i=1}^{n} \bar{a}^{a_i} \leq \prod_{i=1}^{n} a_i^{a_i}$$

(see, e.g., Bernstein [Ber09, Fact 1.18.29]).

Obviously, Theorem 4.42 implies a corresponding result for matrices with nonnegative row or column sums. Since this does not rely on symmetry, we state this result at a later point.

4.7.2 An Improved Inequality Implied by Logarithmic Means

For a more compact notation, we use d_v instead of $d(v)$ for the degree of vertex v in this subsection. The logarithmic mean of two positive real numbers x and y is defined as $(x-y)/(\ln x - \ln y)$, see Carlson [Car72]. It is known that the logarithmic mean is sandwiched between the arithmetic and the geometric mean (see references in [Car72]), that is,

$$\mathfrak{G}(\{x,y\}) = \sqrt{xy} \le \frac{x-y}{\ln x - \ln y} \le \mathfrak{A}(\{x,y\}) = \frac{x+y}{2} \ .$$

This implies for the degrees of two vertices v and u that

$$\frac{2}{(d_u + d_v)} \le \frac{\ln d_v - \ln d_u}{d_v - d_u} \le \frac{1}{\sqrt{d_u d_v}}$$

or

$$\frac{2(d_v - d_u)^2}{(d_u + d_v)} \le (\ln d_v - \ln d_u)(d_v - d_u) \le \frac{(d_v - d_u)^2}{\sqrt{d_u d_v}} \ .$$

Theorem 4.43
For every undirected graph $G = (V, E)$ with average degree $\bar{d}$, we have

$$\sqrt[n]{\sum_{u \in V} \sum_{v \in V} \frac{[d(v) - d(u)]^2}{d_u + d_v}} \left(\prod_{v \in V} d_v \right)^{\bar{d}} \le \prod_{v \in V} d(v)^{d(v)} \ .$$

Proof

$$\sqrt[n]{\sum_{u\in V}\sum_{v\in V}\frac{(d_v-d_u)^2}{(d_u+d_v)}\left(\prod_{v\in V}d_v\right)^{2m/n}} \leq \prod_{\{u,v\}\in E}d_u\cdot d_v$$

$$\sqrt[n]{\sqrt[m]{\sum_{u\in V}\sum_{v\in V}\frac{(d_v-d_u)^2}{(d_u+d_v)}}\prod_{v\in V}d_v^2} \leq \sqrt[m]{\prod_{\{u,v\}\in E}d_u\cdot d_v}$$

$$\sqrt[mn]{\sum_{u\in V}\sum_{v\in V}\frac{(d_v-d_u)^2}{(d_u+d_v)}}\sqrt[n]{\prod_{v\in V}d_v^2} \leq \sqrt[m]{\prod_{\{u,v\}\in E}d_u\cdot d_v}$$

$$\ln\left(\sqrt[mn]{\sum_{u\in V}\sum_{v\in V}\frac{(d_v-d_u)^2}{(d_u+d_v)}}\sqrt[n]{\prod_{v\in V}d_v^2}\right) \leq \ln\sqrt[m]{\prod_{v\in V}d_v^{d_v}}$$

$$\ln\sqrt[mn]{\sum_{u\in V}\sum_{v\in V}\frac{(d_v-d_u)^2}{(d_u+d_v)}}+\ln\sqrt[n]{\prod_{v\in V}d_v^2} \leq \ln\sqrt[m]{\prod_{v\in V}d_v^{d_v}}$$

$$\frac{1}{mn}\sum_{u\in V}\sum_{v\in V}\frac{(d_v-d_u)^2}{(d_u+d_v)}+\frac{2}{n}\sum_{v\in V}\ln d_v \leq \frac{1}{m}\sum_{v\in V}d_v\ln d_v$$

$$\sum_{u\in V}\sum_{v\in V}\frac{(d_v-d_u)^2}{(d_u+d_v)}+\sum_{u\in V}d_u\cdot\sum_{v\in V}\ln d_v \leq \sum_{u\in V}1\cdot\sum_{v\in V}d_v\ln d_v$$

$$\sum_{\{u,v\}\in\binom{V}{2}}\frac{2(d_v-d_u)^2}{(d_u+d_v)} \leq \sum_{\{u,v\}\in\binom{V}{2}}d_v\ln d_v-d_u\ln d_v+d_u\ln d_u-d_v\ln d_u$$

$$\sum_{\{u,v\}\in\binom{V}{2}}\frac{2(d_v-d_u)^2}{(d_u+d_v)} \leq \sum_{\{u,v\}\in\binom{V}{2}}(d_v-d_u)(\ln d_v-\ln d_u)$$

Now the proof is completed by the properties of the logarithmic means.

If the left hand side of the last line in the calculation is replaced by 0, one would obtain the corollary that corresponds to the result of Eliasi and Vukičević [EV13]. Since the term on the left hand side is nonnegative, we obtain a stronger bound.

4.7.3 An Arithmetic-Geometric-Harmonic Mean Inequality

$H(G):=\sum_{\{u,v\}\in E}\frac{2}{d_u+d_v}$ is called the *harmonic index* of the graph G. It was shown by Ilić [Ili12] and Xu [Xu12] that

$$H(G)\geq\frac{2m^2}{M_1(G)}\quad.$$

We show the following refinement of this inequality using another multiplicative variant of the first Zagreb index which was considered by Eliasi, Iranmanesh and Gutman [EIG12]:

$$\Pi_1^*=\prod_{\{u,v\}\in E}(d_u+d_v)\quad.$$

Theorem 4.44

For every undirected graph with m edges, we have

$$\frac{w_1}{H(G)} = \frac{2m}{H(G)} \leq \sqrt[m]{\Pi_1^*} \leq \frac{M_1}{m} = \frac{2w_2}{w_1} \ .$$

Proof   The inequality of arithmetic, geometric, and harmonic means directly implies

$$\frac{m}{\sum_{\{u,v\}\in E} \frac{1}{d_u+d_v}} \leq \sqrt[m]{\prod_{\{u,v\}\in E} (d_u + d_v)} \leq \frac{1}{m} \sum_{\{u,v\}\in E} (d_u + d_v) \ .$$

The sum on the right hand side corresponds to $M_1 = w_2$. Apart from the missing factor of 2, the denominator of the fraction on the left hand side corresponds to $H(G)$. The proof is complete by $w_1 = 2m$.

Chapter 5

Restricted Graph Classes

It is not so much whether a theorem is useful that matters, but how elegant it is.

Stanisław Ulam

Probably unaware of the work by Lagarias, Mazo, Shepp and McKay [LMSM84], the special case $w_0 w_3 \geq w_1 w_2$ was conjectured again by Hansen, but in a slightly different form using the Zagreb indices $M_1 = \sum_{v \in V} d_v^2 = w_2$ and $M_2 = \sum_{\{x,y\} \in E} d_x d_y = w_3/2$. Hansen and Vukičević [HV07] again found counterexamples, but they proved the inequality for graphs with maximum degree not exceeding 4. That includes most of the graphs that are interesting from the perspective of chemical structures. An undirected graph is called a *chemical graph* if its maximum degree Δ is bounded by $\Delta \leq 4$.

Theorem 5.1 (Hansen and Vukičević)
For all chemical graphs with Zagreb indices M_1 and M_2, we have

$$\frac{M_1}{n} \leq \frac{M_2}{m} .$$

The next two sections recapitulate findings from the article [TWK$^+$13].

5.1 Counterexamples for Special Cases

5.1.1 Bipartite Graphs

We show that bipartite graphs violate the inequality $w_r \cdot w_s \leq n \cdot w_{r+s}$, in particular for $r = 2$, $s = 1$. Similar to the general counterexamples proposed by Lagarias et al., our counterexamples consist of two parts: a star and (instead of complete graphs) complete bipartite graphs. The numbers of walks are:

$$
\begin{aligned}
w_0(B_{n/2,n/2}) &= n & w_0(S_n) &= n \\
w_1(B_{n/2,n/2}) &= n \cdot \left(\tfrac{n}{2}\right) & w_1(S_n) &= 2(n-1) \\
w_2(B_{n/2,n/2}) &= n \cdot \left(\tfrac{n}{2}\right)^2 & w_2(S_n) &= n(n-1) \\
w_3(B_{n/2,n/2}) &= n \cdot \left(\tfrac{n}{2}\right)^3 & w_3(S_n) &= 2(n-1)^2
\end{aligned}
$$

Consider for instance the graph consisting of the complete bipartite graph $B_{2,2}$ and the star S_6. For this graph, we have $w_0 = 4+6$, $w_1 = 8+10$, $w_2 = 16+30$, and $w_3 = 32+50$. Hence, the inequality is violated: $w_1w_2 = 828 \not\leq 820 = w_0w_3$.

Connected counterexamples can be constructed by appropriate scaling and attaching both parts through a single edge. Consider a graph $G = (V, E)$ consisting of a complete bipartite graph $B_{x,x}$ and a star S_y, where an extra edge is added between one of the leaves of the star and one of the vertices of the $B_{x,x}$. Then we have

$$
\begin{aligned}
w_0 &= |V| = 2x + y \\
w_1 &= 2|E| = 2(x^2 + y) \\
w_2 &= \sum_{v \in V} d(v)^2 \\
&= (2x-1)x^2 + (x+1)^2 + (y-1)^2 + 2^2 + (y-2) \cdot 1^2 \\
&= 2x^3 - x^2 + x^2 + 2x + 1 + y^2 - 2y + 1 + 4 + y - 2 \\
&= 2x^3 + 2x + y^2 - y + 4 \\
w_3 &= \sum_{\{v,w\} \in E} d(v) \cdot d(w) \\
&= 2[x(x-1)x^2 + x(x+1)x + 2(x+1) + 2(y-1) + (y-2)(y-1)] \\
&= 2[x^4 - x^3 + x^3 + x^2 + 2x + 2 + 2y - 2 + y^2 - 3y + 2] \\
&= 2[x^4 + x^2 + 2x + y^2 - y + 2]
\end{aligned}
$$

Thus, we ask whether the following inequality is valid:

$$
\begin{aligned}
w_1 w_2 &\overset{?}{\leq} w_0 w_3 \\
2(x^2 + y) \cdot (2x^3 + 2x + y^2 - y + 4) &\leq (2x + y) \cdot 2(x^4 + x^2 + 2x + y^2 - y + 2) \\
\begin{array}{l} 2x^5 + 2x^3 + x^2y^2 - x^2y + 4x^2 \\ +2x^3y + 2xy + y^3 - y^2 + 4y \end{array} &\leq \begin{array}{l} 2x^5 + 2x^3 + 4x^2 + 2xy^2 - 2xy + 4x \\ +x^4y + x^2y + 2xy + y^3 - y^2 + 2y \end{array} \\
x^2y^2 - x^2y + 2x^3y + 4y &\leq 2xy^2 - 2xy + 4x + x^4y + x^2y + 2y \\
x^2y^2 + 2x^3y + 2y + 2xy &\leq 2xy^2 + 4x + x^4y + 2x^2y
\end{aligned}
$$

Now, trying $x = 3$ yields

$$9y^2 + 54y + 2y + 6y \overset{?}{\leq} 6y^2 + 12 + 81y + 18y$$
$$3y^2 \leq 12 + 37y$$

and this is *not valid* for $y \geq 13$.

This also yields a smaller connected counterexample (using 19 vertices) than the one in [LMSM84].

5.1.2 Forests and Trees

Now we show that there are arbitrarily large cycle-free graphs (forests) contradicting the inequality. Again, these graphs consist of two parts. This time, the two compounds are a path and a star. The respective numbers of walks for the path (assume $n \geq 3$) are $w_0(P_n) = n$, $w_1(P_n) = 2(n-1) = 2n - 2$, $w_2(P_n) = (n-2) \cdot 2^2 + 2 \cdot 1^2 = 4n - 6$ (for $n \geq 2$), $w_3(P_n) = 2[(n-3) \cdot 2 \cdot 2 + 2 \cdot 2 \cdot 1] = 8n - 16$ (for $n \geq 3$). For the star, we have $w_0(S_n) = n$, $w_1(S_n) = 2(n-1)$, $w_2(S_n) = n(n-1)$, and $w_3(S_n) = 2(n-1)^2$. Now consider a graph consisting of a star S_x and a path P_y. Then the inequality reads as follows:

$$
\begin{aligned}
(x+y)[2 \cdot (x-1)^2 + 8y - 16] &\geq [x(x-1) + 4y - 6][2(x-1) + 2(y-1)] \\
(x+y)((x-1)^2 + 4y - 8) &\geq (x(x-1) + 4y - 6)(x + y - 2) \\
(x+y)(x^2 - 2x + 4y - 7) &\geq (x^2 - x + 4y - 6)(x + y - 2) \\
\begin{array}{l} x^3 - 2x^2 + 4xy - 7x \\ +x^2 y - 2xy + 4y^2 - 7y \end{array} &\geq \begin{array}{l} x^3 - x^2 + 4xy - 6x + x^2 y - xy \\ +4y^2 - 6y - 2x^2 + 2x - 8y + 12 \end{array} \\
x^2 - 3x - xy + 7y - 12 &\geq 0
\end{aligned}
$$

Values for x from 2 to 7 result in inequalities that are true for $y > 2$, but already setting $x = 8$ leads to $-y + 28 \geq 0$ which is not valid for $y \geq 29$. Thus, a possible counterexample consists of the star S_8 and the path P_{29}.

Surprisingly, the connected variant is no longer a counterexample. As we will see in the next section, the inequality $w_1 w_2 \leq w_0 w_3$ is valid for all trees (see Theorem 5.3).

5.1.3 Construction of Worst Case Trees

In order to answer the question whether the inequality $w_1 w_2 \leq w_0 w_3$ holds for all trees we investigate the behavior of different trees with respect to the value of the difference of both sides $(w_0 w_3 - w_1 w_2)$. Within this subsection, we will show how to construct trees of a given degree sequence that minimize this difference (i.e., "worst case trees"). Later on, our aim is to show that certain graph transformations change the value of the difference in a certain direction which leads to a proof of the inequality.

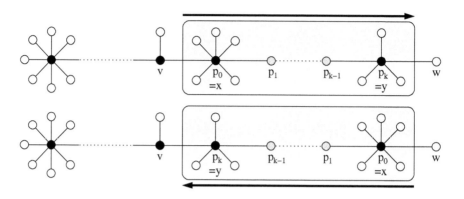

Figure 5.1: Path inversion in a (caterpillar) tree.

Lemma 5.1

For a given degree sequence of a tree, the tree that minimizes $w_0 w_3 - w_1 w_2$ cannot have four different vertices $v, w, x, y \in V$ with $d(x) > d(y)$ and $d(v) > d(w)$ such that x and y are the neighbors of v and w, respectively, on the path from v to w.

Proof   Assume the contrary, i.e., there is a worst case tree having minimum difference value for a given degree sequence that has such vertices $v, w, x, y \in V$ (see Figure 5.1). Consider the tree that is constructed by inverting the x-y-path between v and w (i.e., x is now connected to the former neighbor w of y, whereas y's connection to w is replaced by the connection to the former neighbor v of x). This tree has the same degree sequence as before, i.e., besides the number of vertices n and the number of edges $m = w_1/2$ also the total number of 2-step walks $w_2 = \sum_{v \in V} d(v)^2$ has not changed. For the number of 3-step walks $w_3 = 2 \sum_{\{s,t\} \in E} d(s)d(t)$ only the values for the edges connecting the x-y-path to v and w have changed from $d(x)d(v) + d(y)d(w)$ to $d(y)d(v) + d(x)d(w)$.

$$d(x)d(v) + d(y)d(w) > d(y)d(v) + d(x)d(w)$$
$$[d(x) - d(y)]\, d_v - [d(x) - d(y)]\, d_w > 0$$
$$[d(x) - d(y)]\, [d(v) - d(w)] > 0$$

Since $d(x) > d(y)$ and $d(v) > d(w)$, the value of w_3 must have become smaller, a contradiction to the assumption that $w_0 w_3 - w_1 w_2$ was a minimum.

At first, we have a look at a special class of trees, namely the caterpillar trees. A caterpillar is a tree that has all its leaves attached to a central path. For a given degree sequence of a caterpillar, a caterpillar that minimizes the value of the difference $w_0 w_3 - w_2 w_1$ has a vertex of maximum degree as one of the end vertices of its central path. This is a direct consequence of the lemma. Furthermore, the other end vertex of the central path must be the second vertex in the order of nonincreasing degrees. (Note that there may be more than one caterpillar tree topology minimizing the difference value in the case where a vertex degree > 1 occurs more than just once.) The next two vertices towards the inside of the central path must be two of the remaining vertices with lowest possible degree.

The lemma directly implies an algorithm for the construction of a worst-case caterpillar (i.e., a caterpillar that minimizes $w_0 w_3 - w_2 w_1$): From the given degree sequence, we start with the two leaf-ends of the central path (with minimum degree 1) and fill in the remaining vertices from the outside to the middle by alternately considering two remaining vertices of maximum or minimum degree, starting with the two vertices of maximum degree, followed by the two remaining vertices of minimum degree and so on. We only have to make sure that, if the two vertices inserted in the last iteration differ in their degree and also the two vertices to be inserted in the current iteration differ in their degree, then the higher-degree vertex of one pair must get the edge to the lower-degree vertex of the other pair and vice versa. The result is a caterpillar that has its vertices of most extreme degrees at the ends of the central path, minimum and maximum alternating towards the center, and the vertices corresponding to the median of the degree sequence are located in the center.

Now we consider arbitrary trees. The lemma implies that in a worst-case tree for a given degree sequence, a vertex x of maximum degree cannot have more than one neighboring inner vertex while at the same time there exists a vertex y with lower degree that has a neighboring leaf w. (Otherwise there is a non-leaf neighbor v of x that is not on the path from x to y and the lemma could be applied since $d_v \geq 2 > d_w = 1$ and $d_x > d_y$.) The lemma not only implies that the vertices of maximum degrees must have as many neighboring leaves as possible, it also implies as a next step that if there is a non-leaf (inner) neighbor of such a vertex, this vertex must have smallest possible degree. Hence, we can build a worst-case tree from a given degree sequence from the outside to the inside. The outer shell is the set of leaves, the next layers towards the inside of the tree are made of vertices having largest and smallest possible degree in an alternating fashion. Only one of the valences has to be left for attaching this subtree to the rest of the graph. (Note that there may be several worst-case trees with different topologies if there are vertices having the same degree.) A similar algorithm has been proposed by Wang [Wan08] for a related problem.

5.2   Semiregular Graphs

Assume that the graph is bipartite and there are n_1 vertices having degree d_1 on one side of the partition and n_2 vertices having degree d_2 on the other side. Such a graph is called *semiregular*. Then we have $n = n_1 + n_2$ vertices and $m = (n_1 d_1 + n_2 d_2)/2$ edges. We also know that $n_1 d_1 = n_2 d_2$. Furthermore, the walk numbers are easily calculated as

$$
\begin{aligned}
w_{2k} &= (n_1 + n_2) d_1^k d_2^k \\
w_{2k+1} &= n_1 d_1^{k+1} d_2^k + n_2 d_1^k d_2^{k+1} = (n_1 d_1 + n_2 d_2) d_1^k d_2^k = 2m d_1^k d_2^k \ ,
\end{aligned}
$$

see also Nikiforov [Nik06b]. In a recent discussion, Tamás Réti[1] proposed the equality $w_0 w_{2k+1} = w_1 w_{2k}$ for all semiregular graphs. In fact, we can state this more generally as follows.

---

[1] Private communication.

Theorem 5.2
For every semiregular graph G and a, b, c, d ∈ ℕ with a + b = c + d, we have

$$
\begin{aligned}
w_{2a}(G)w_{2b}(G) &= w_{2c}(G)w_{2d}(G) \\
w_{2a+1}(G)w_{2b+1}(G) &= w_{2c+1}(G)w_{2d+1}(G) \\
w_{2a}(G)w_{2b+1}(G) &= w_{2c}(G)w_{2d+1}(G) \\
w_{2a+1}(G)w_{2b+1}(G) &\leq w_{2c}(G)w_{2d+2}(G) \; .
\end{aligned}
$$

Proof All three equalities follow straight from applying the formula above for the walk numbers of odd and even lengths in semiregular graphs. For the inequality, we obtain

$$
\begin{aligned}
(n_1 d_1 + n_2 d_2)^2 d_1^{a+b} d_2^{a+b} &\leq (n_1 + n_2)^2 d_1^{c+d+1} d_2^{c+d+1} \\
(n_1 d_1 + n_2 d_2)^2 &\leq (n_1 + n_2)^2 d_1 d_2 \quad \text{(since } a + b = c + d) \\
w_1(G)^2 &\leq w_0(G)w_2(G) \; .
\end{aligned}
$$

The last inequality follows from Theorem 4.3.

Note that the semiregular graphs include the complete bipartite graphs (in particular the stars) and the subdivision graphs of regular graphs (where each edge is split into two edges).

5.3 Trees

For a compact notation, we will often use d_v instead of $d(v)$ for the degree of vertex v within this section.

5.3.1 Stars and Paths

From the previous section, we can deduce that a star consisting of n vertices and $m = n - 1$ edges has $w_{2k} = n(n-1)^k$ and $w_{2k+1} = 2(n-1)^{k+1}$ for all $k \in \mathbb{N}$.

Since stars are complete bipartite graphs, we obtain the following result from Theorem 5.2.

Corollary 5.1
For every star S and a, b, c, d ∈ ℕ with a + b = c + d, we have

$$
\begin{aligned}
w_{2a}(S)w_{2b}(S) &= w_{2c}(S)w_{2d}(S) \\
w_{2a+1}(S)w_{2b+1}(S) &= w_{2c+1}(S)w_{2d+1}(S) \\
w_{2a}(S)w_{2b+1}(S) &= w_{2c}(S)w_{2d+1}(S) \\
w_{2a+1}(S)w_{2b+1}(S) &\leq w_{2c}(S)w_{2d+2}(S) \; .
\end{aligned}
$$

Accordingly, the most general conceivable sandwich theorem $w_{a+b} \cdot w_{a+c} \le w_a \cdot w_{a+b+c}$ is not valid even for the very restricted class of trees, since we can easily construct stars with $w_2 \cdot w_2 > w_1 \cdot w_3$ (or $\frac{w_3}{w_2} < \frac{w_2}{w_1}$).

Lemma 5.2
For every path P_n with n vertices, we have

$$w_1 \cdot w_k \le w_0 \cdot w_{k+1} \ .$$

Proof   Suppose that $G = (V, E) = P_n$ is a path with $n \ge 1$ vertices. Let b denote a leaf of P and $k \in \mathbb{N}$. Then

$$
\begin{aligned}
w_0 w_{k+1} - w_1 w_k &= n w_{k+1} - 2(n-1)w_k \\
(w_0 w_{k+1} - w_1 w_k)/n &= (n w_{k+1} - 2n w_k + 2 w_k)/n \\
&= \sum_{v \in V} d_v w_k(v) - 2\left(\sum_{v \in V} w_k(v)\right) + 2 w_k/n \\
&= \sum_{v \in V} (d_v - 2) w_k(v) + 2 w_k/n \ = \ -2 w_k(b) + 2 w_k/n \ .
\end{aligned}
$$

The last equality follows from the fact that each inner vertex v of a path has degree $d_v = 2$ and each leaf b has degree $d_b = 1$. Now we show $2 w_k/n - 2 w_k(b) \ge 0$ by proving $w_k(v) \ge w_k(b)$ for every vertex v.

Case 1:

The distance between v and b is even. For each walk starting at b, we construct a unique walk starting at v by symmetrically mimicking all steps until both walks meet at the same vertex. After that, the constructed walk uses the same edges as the walk that started at b.

Case 2:

The distance between v and b is odd. If v is the other leaf then we are done (because of symmetry). Thus, assume that v is not a leaf. Now, we construct the corresponding walk in nearly the same way as in the first case, but we ignore the first step which is fixed anyways. Now the distance to v is even and we apply the same method as in the first case (which is possible since v is not a leaf). After that, the last move can be chosen arbitrarily.

5.3.2   Walks of Length 2 and 3 in Trees

Let p_i denote the number of directed paths (i.e., vertex-disjoint walks) of length i in the graph. Further, let $w_{i,j}$ and $w_{i,j}(v)$ denote the number of walks (total or starting at $v \in V$, respectively) having length i, where the last j edges constitute a *path*.

Lemma 5.3
For every undirected graph and $2 \le i \in \mathbb{N}$, the equality $w_i(v) = w_{i,2}(v) + w_{i-1}(v)$ holds for all $v \in V$. Thus, we have $w_i = w_{i,2} + w_{i-1}$.

Proof   Each walk of length $i - 1$ can be extended to a walk of length i just by traversing the last edge in the opposite direction. In all other walks of length i, the last two steps do not traverse the same edge, i.e., they end with a path of length 2.

Theorem 5.3
For every tree, the following inequality is valid:

$$w_1 \cdot w_2 \leq w_0 \cdot w_3 \qquad \text{or, equivalently,} \qquad \overline{d} \cdot w_2 \leq w_3 \ .$$

Proof   Besides $w_0 = n$, we have $m = (n - 1)$ for every tree, and thus we know that $w_1 = 2m = 2(n - 1)$. By Lemma 5.3, we have

$$
\begin{aligned}
w_2 &= w_{2,2} + w_1 = p_2 + w_1 \qquad \text{and} \\
w_3 &= w_{3,2} + w_2 = p_3 + p_2 + w_2 = p_3 + 2p_2 + w_1 \ .
\end{aligned}
$$

Consider the difference of both sides of the inequality:

$$
\begin{aligned}
& w_0 w_3 && - w_1 w_2 && \\
=\ & n w_3 && - 2(n-1) w_2 && \\
=\ & n[w_3 && - 2 w_2] && + 2 w_2 \\
=\ & n[(w_{3,2} + w_2) && - 2 w_2] && + 2 w_2 \\
=\ & n[w_{3,2} && - w_2] && + 2 w_2 \\
=\ & n[(p_3 + p_2) && - (p_2 + w_1)] && + 2(p_2 + w_1) \\
=\ & n[p_3 && - p_1] && + 2(p_2 + p_1) \\
=\ & n[p_3 && - 2(n-1)] && + 2(p_2 + 2(n-1)) \\
=\ & n[p_3 && - 2n + 6] && + 2(p_2 - 2)
\end{aligned}
$$

Note that each tree G with diameter $\operatorname{diam}(G) \leq 2$ is a star. In this case, we have $w_0 w_3 = w_1 w_2$ (see Corollary 5.1).

Let $G = (V, E)$ be any tree that satisfies the conditions $\operatorname{diam}(G) \geq 3$, $(p_3 - 2n + 6) \geq 0$, and $w_1 w_2 \leq w_0 w_3$. Then, we can create a new tree G' by appending a leaf to an arbitrary vertex. For G', we have $n' = n + 1$, $p_2' \geq p_2 + 2$, and $p_3' \geq p_3 + 2$. Hence, G' satisfies the three conditions, too.

Each tree having diameter at least 3 can be constructed by repeatedly appending new leaves to a path of length 3. For the path of length 3, we have $n = 4$, $p_2 = 4$, and $p_3 = 2$. Hence, all conditions are satisfied, and therefore, all trees observe the above inequality.

After the publication of this proof [HKM+12], we found out that essentially the same result had already been shown by Vukičević and Graovac [VG07] (and later again by Andova, Cohen and Škrekovski [ACŠ12]) in the same form as Theorem 5.1, that is, as an inequality for the Zagreb indices.

Theorem 5.4 (Vukičević and Graovac)
Let T be a tree with $n \geq 2$ vertices and m edges. Then,

$$\frac{M_1(T)}{n} \leq \frac{M_2(T)}{m} .$$

The equality holds if and only if T is a star.

Hence, our proof is just a short alternative to show this result. We will use a similar proof technique in the next subsection. Maybe this concept could lead to a more general result which is stated as a conjecture at the end of this section.

5.3.3   Walks of Length 4 and 5 in Trees

Theorem 5.5
For every tree, the following inequality is valid:

$$w_1 \cdot w_4 \leq w_0 \cdot w_5 \qquad \text{or, equivalently,} \qquad \overline{d} \cdot w_4 \leq w_5 .$$

The proof of this theorem consists of two parts. In the first part, we show the inequality for a subset of trees having diameter at most 4 (barbell graphs). In the second part, we use induction in a similar way as in the proof of Theorem 5.3 to show the inequality for all remaining trees.

Let $G = (V, E)$ be a tree. Then for each $i \in \mathbb{N}$, the following inequalities are equivalent:

$$w_0 w_{i+1} \geq w_1 w_i \qquad \Leftrightarrow \qquad n w_{i+1} \geq 2(n-1) w_i$$

$$\Leftrightarrow \qquad 2 w_i + n \sum_{v \in V} d_v w_i(v) - 2n \sum_{v \in V} w_i(v) \geq 0$$

$$\Leftrightarrow \qquad 2 w_i / n + \sum_{v \in V} (d_v - 2) w_i(v) \geq 0 .$$

Thus, for a proof of the inequality, every i-step walk that starts at a leaf creates a negative unit that has to be compensated for by the contribution of the respective neighbors and the correction term $2 w_i / n$.

Let b be a leaf attached at an inner vertex x. Then, for $i > 0$, we have $w_i(x) = w_{i+1}(b)$, and using Lemma 5.3 we obtain

$$(d_x - 2) w_{i+1}(x) = (d_x - 2)(w_i(x) + w_{i+1,2}(x)) \geq (d_x - 2) w_{i+1}(b) .$$

So we can use the positive units of an inner vertex x to compensate for the negative units of at least $d_x - 2$ attached leaves. This is called *deficit adjustment* at vertex x.

5.3.3.1 $(1, n_1, n_2)$-Barbell Graphs (Trees with Diameter 3)

Suppose that an (ℓ, n_1, n_2)-*barbell graph* is a graph that consists of a path of length ℓ, having attached $n_1 \geq 1$ and $n_2 \geq n_1$ leaves at the two end vertices x_1 and x_2, respectively. Each $(1, n_1, n_2)$-barbell graph is a tree having diameter 3 and every tree with diameter 3 is a $(1, n_1, n_2)$-barbell graph for properly chosen n_1, n_2. In the following, we show $w_0 w_{i+1} \geq w_1 w_i$ for each $i \in \mathbb{N}$ and every $(1, n_1, n_2)$-barbell graph (and thus for all trees having diameter 3).

Lemma 5.4
For all $(1, n_1, n_2)$-barbell graphs and $i \in \mathbb{N}$, we have $w_i(x_1) \leq w_i(x_2)$.

Proof   We will prove the inequality by induction over $i \in \mathbb{N}$. For $i = 0$ and $i = 1$, the inequality holds since $w_0(x_1) = 1 = w_0(x_2)$ and $w_1(x_1) = n_1 + 1 \leq n_2 + 1 = w_1(x_2)$.
Let $i \geq 2$. Each of the walks of length i starting at x_1 could go to x_2 or to one of the n_1 leaves in the first step. Since each of those leaves has the same number of walks of length $i - 1$ and all of these walks visit x_1 in the next step, we obtain

$$w_i(x_1) = w_{i-1}(x_2) + n_1 w_{i-2}(x_1) \ .$$

From Lemma 5.3, we know that $w_i(x_2) = w_{i-1}(x_2) + w_{i,2}(x_2)$.
Now, we distinguish even and odd values for the length i. If i is even then each walk of length $i - 2$ must end either at x_2 or at one of the n_1 leaves of x_1. Each of those walks can be extended by exactly n_1 directed paths of length 2. Similarly, for odd i, any walk of length $i - 2$ can be extended by exactly n_2 directed paths of length 2. Then we have

$$w_{i,2}(x_2) = \begin{cases} n_1 w_{i-2}(x_2) & \text{if } i \text{ is even,} \\ n_2 w_{i-2}(x_2) & \text{if } i \text{ is odd.} \end{cases}$$

Assume that $w_j(x_1) \leq w_j(x_2)$ holds for all $j < i$.

$$w_i(x_1) = w_{i-1}(x_2) + n_1 w_{i-2}(x_1) \leq \begin{cases} w_{i-1}(x_2) + n_1 w_{i-2}(x_2) = w_i(x_2) & \text{if } i \text{ is even,} \\ w_{i-1}(x_2) + n_2 w_{i-2}(x_2) = w_i(x_2) & \text{if } i \text{ is odd.} \end{cases}$$

The lemma follows by induction.

Lemma 5.5
For all $(1, n_1, n_2)$-barbell graphs and $i \in \mathbb{N}$, we have $w_0 w_{i+1} \geq w_1 w_i$.

Proof   For $i = 0$, both sides are equal, thus assume $i > 0$. Let b_1 and b_2 be leaves attached to x_1 and x_2, respectively. We perform a deficit adjustment at the vertices x_1 and x_2. There are $(d_{x_1} - 1)$ and $(d_{x_2} - 1)$ leaves attached at x_1 and x_2. $(d_{x_1} - 2)$ and $(d_{x_2} - 2)$ of them can be compensated for by the excess terms of x_1 and x_2. Hence, at most $w_i(b_1) + w_i(b_2) = w_{i-1}(x_1) + w_{i-1}(x_2)$ negative units remain unbalanced.

Now, we show $2w_i/n \geq w_{i-1}(x_1) + w_{i-1}(x_2)$. By Lemma 5.3, we know that $w_{i-1}(x_1) \leq w_i(x_1)$ and $w_{i-1}(x_2) \leq w_i(x_2)$. Lemma 5.4 implies $w_{i-1}(x_1) \leq w_{i-1}(x_2)$, and since $n_1 \leq n_2$ we obtain $n_1 w_{i-1}(x_1) + n_2 w_{i-1}(x_2) \geq n_1 w_{i-1}(x_2) + n_2 w_{i-1}(x_1)$. Therefore, we have

$$
\begin{aligned}
2w_i/n &= 2[n_1 w_i(b_1) + w_i(x_1) + n_2 w_i(b_2) + w_i(x_2)]/n \\
&= 2[n_1 w_{i-1}(x_1) + w_i(x_1) + n_2 w_{i-1}(x_2) + w_i(x_2)]/n \\
&\geq [n_1 w_{i-1}(x_1) + n_2 w_{i-1}(x_1) + 2w_i(x_1) \\
&\quad + n_1 w_{i-1}(x_2) + n_2 w_{i-1}(x_2) + 2w_i(x_2)]/n \\
&\geq [n_1 w_{i-1}(x_1) + n_2 w_{i-1}(x_1) + 2w_{i-1}(x_1) \\
&\quad + n_1 w_{i-1}(x_2) + n_2 w_{i-1}(x_2) + 2w_{i-1}(x_2)]/n \\
&\geq [n w_{i-1}(x_1) + n w_{i-1}(x_2)]/n = w_{i-1}(x_1) + w_{i-1}(x_2).
\end{aligned}
$$

This completes the proof.

5.3.3.2 $(2, n_1, n_2)$-Barbell Graphs

Lemma 5.6
For all $(2, n_1, n_2)$-barbell graphs and $i \in \mathbb{N}$, we have $w_i(x_1) \leq w_i(x_2)$.

Proof   For each walk starting at x_1, we can construct a unique walk of the same length that starts at x_2: Since $n_1 \leq n_2$, we can injectively map each leaf of x_1 to a leaf of x_2. For each walk starting at x_1, we mimic this walk (using the mapping) until the walk passes the center. From this point on, we follow exactly the same way (without using the mapping).

Lemma 5.7
For all $(2, n_1, n_2)$-barbell graphs and $i \in \mathbb{N}$, we have $n w_{i+1} \geq 2(n-1)w_i$.

Proof   Let b_1 and b_2 be leaves attached to x_1 and x_2, respectively. We perform a deficit adjustment at the vertices x_1 and x_2. Hence, at most $w_i(b_1)+w_i(b_2) = w_{i-1}(x_1)+w_{i-1}(x_2)$ negative units remain unbalanced. Now, we show $2w_i/n \geq w_{i-1}(x_1)+w_{i-1}(x_2)$. By Lemma 5.3, we know that $w_{i-1}(x_1) \leq w_i(x_1)$ and $w_{i-1}(x_2) \leq w_i(x_2)$. Furthermore, the central node c satisfies the equality $w_i(c) = w_{i-1}(x_1) + w_{i-1}(x_2)$. Now, from $n_1 \leq n_2$ and Lemma 5.6 we obtain

$$
\begin{aligned}
2w_i/n &= 2[n_1 w_i(b_1) + w_i(x_1) + n_2 w_i(b_2) + w_i(x_2) + w_i(c)]/n \\
&= 2[n_1 w_{i-1}(x_1) + w_i(x_1) + n_2 w_{i-1}(x_2) + w_i(x_2) + w_{i-1}(x_1) + w_{i-1}(x_2)]/n \\
&= 2[(n_1 + 1)w_{i-1}(x_1) + w_{i-1}(x_1) + w_{i,2}(x_1) \\
&\quad +(n_2 + 1)w_{i-1}(x_2) + w_{i-1}(x_2) + w_{i,2}(x_2)]/n \\
&= 2[(n_1 + 2)w_{i-1}(x_1) + w_{i,2}(x_1) + (n_2 + 2)w_{i-1}(x_2) + w_{i,2}(x_2)]/n \\
&\geq [(n_1 + n_2 + 4)w_{i-1}(x_1) + 2w_{i,2}(x_1) + (n_1 + n_2 + 4)w_{i-1}(x_2) + 2w_{i,2}(x_2)]/n \\
&\geq [(n_1 + n_2 + 3)w_{i-1}(x_1) + (n_1 + n_2 + 3)w_{i-1}(x_2)]/n \\
&\geq [n w_{i-1}(x_1) + n w_{i-1}(x_2)]/n \; = \; w_{i-1}(x_1) + w_{i-1}(x_2).
\end{aligned}
$$

This completes the proof.

5.3.3.3   *Proof for the w_5-Inequality for Trees*

We use $N_i(v)$ to denote the number of vertices having distance i from v. Besides $w_1(v) = d_v$, $w_2(v) = d_v + N_2(v)$, and $w_3(v) = d_v^2 + N_2(v) + N_3(v)$, we observe the following equalities for trees:

$$
\sum_{v \in V}(d_v - 1)N_i(v) = \sum_{v \in V} N_{i+1}(v) = p_{i+1} \quad \text{or} \quad \sum_{v \in V} d_v N_i(v) = p_i + p_{i+1}
$$

$$
\sum_{v \in V} N_2(v)^2 = p_2 + p_4 + \sum_{v \in V} d_v(d_v - 1)(d_v - 2)
$$

$$
\sum_{v \in V} N_2(v)N_3(v) = p_3 + p_5 + \sum_{v \in V} N_2(v)(d_v - 1)(d_v - 2)
$$

$$w_4 = \sum_{v \in V} w_2(v)^2 = \sum_{v \in V} [d_v + N_2(v)]^2 = \sum_{v \in V} d_v^2 + 2d_v N_2(v) + N_2(v)^2$$

$$= w_2 + 3p_2 + 2p_3 + p_4 + \sum_{v \in V} d_v(d_v - 1)(d_v - 2)$$

$$w_4 = \sum_{v \in V} w_1(v)w_3(v) = \sum_{v \in V} d_v \cdot [d_v^2 + N_2(v) + N_3(v)]$$

$$= \left(\sum_{v \in V} d_v^3\right) + p_2 + 2p_3 + p_4$$

$$w_5 = \sum_{v \in V} w_2(v)w_3(v) = \sum_{v \in V} [d_v + N_2(v)] \cdot [d_v^2 + N_2(v) + N_3(v)]$$

$$= \sum_{v \in V} d_v^3 + d_v N_2(v) + d_v N_3(v) + d_v^2 N_2(v) + N_2(v)^2 + N_2(v)N_3(v)$$

$$= \left(\sum_{v \in V} d_v^3\right) + p_2 + p_3 + p_3 + p_4 + \left(\sum_{v \in V} d_v^2 N_2(v)\right) + p_2 + p_4$$

$$+ \left(\sum_{v \in V} d_v(d_v - 1)(d_v - 2)\right) + p_3 + p_5 + \left(\sum_{v \in V} N_2(v)(d_v - 1)(d_v - 2)\right)$$

$$= \left(\sum_{v \in V} d_v^3 + d_v^2 N_2(v) + d_v(d_v - 1)(d_v - 2) + N_2(v)(d_v - 1)(d_v - 2)\right)$$

$$+ 2p_2 + 3p_3 + 2p_4 + p_5$$

$$= \left(\sum_{v \in V} (d_v + N_2(v)) \left(d_v^2 + (d_v - 1)(d_v - 2)\right)\right) + 2p_2 + 3p_3 + 2p_4 + p_5$$

Accordingly, this results in

$$w_0 w_5 - w_1 w_4 = n w_5 - 2(n-1)w_4 = n[w_5 - 2w_4] + 2w_4$$

$$= n\left[\left(\sum_{v \in V} (d_v + N_2(v)) \left(d_v^2 + (d_v - 1)(d_v - 2)\right)\right) + 2p_2 + 3p_3 + 2p_4 + p_5\right.$$

$$- w_2 - 3p_2 - 2p_3 - p_4 - \left(\sum_{v \in V} d_v(d_v - 1)(d_v - 2)\right)$$

$$\left. - \left(\sum_{v \in V} d_v^3\right) - p_2 - 2p_3 - p_4\right] + 2w_4$$

$$= n\left[\left(\sum_{v \in V} d_v^2 N_2(v) + N_2(v)(d_v - 1)(d_v - 2)\right) - w_2 - 2p_2 - p_3 + p_5\right] + 2w_4$$

$$= n\left[\left(\sum_{v \in V} N_2(v)d_v(2d_v - 3)\right) - w_2 - p_3 + p_5\right] + 2w_4 \ .$$

Lemma 5.8
Every tree with n vertices and diameter at least 3 has at least $6n$ walks of length 4.

Proof For the path graph of length 3, the inequality $w_4 = 26 > 24 = 6 \cdot 4 = 6n$ holds. Let B be a tree with $\mathrm{diam}(B) \geq 3$. If we attach a leaf b via edge $\{b, x\}$ to B, this leaf is the starting point of a path of length 3. There are 14 walks of length 4 that use only edges of this path and contain the edge $\{b, x\}$. Therefore, every additional vertex introduces at least 14 new walks of length 4.

Since every tree with diameter at least 3 can be constructed by iteratively attaching new leaves to P_4, the lemma follows.

By application of Lemma 5.8, it is sufficient to show

$$w_5 - 2w_4 + 12 = \left(\sum_{v \in V} N_2(v)d_v(2d_v - 3)\right) - w_2 - p_3 + p_5 + 12 \geq 0 \ .$$

We show this inequality by induction for all graphs having diameter at least 4, except for the $(2, n_1, n_2)$-barbell graphs with $n_2 \geq 2$. Each such graph can be constructed from a path of length 4 by iteratively adding leaves in such a way that no intermediate step results in a $(2, n_1, n_2)$-barbell graph with $n_2 \geq 2$. To this end, we observe that a graph with diameter at least 4 contains a path of length 4 with an additional leaf attached to its center or a path of length 5 if and only if it is not a $(2, n_1, n_2)$-barbell graph. Hence, we start the construction by adding a leaf either to the center or to an end vertex of the path of length 4.

For the path graph of length 4, we have $w_5 - 2w_4 + 12 \geq 0$ (since $w_5 = 72$ and $w_4 = 42$).

Now we show that the term $\left(\sum_{v \in V} N_2(v) d_v (2d_v - 3)\right) - w_2 - p_3 + p_5$ cannot decrease by attaching a new leaf, if the graph had diameter at least 4 before and is not a $(2, n_1, n_2)$-barbell graph with $n_2 \geq 2$.

Let $G = (V, E)$ be the original tree satisfying these requirements, let b denote the new leaf, and let x be the unique vertex adjacent to b. Further, let $G' = (V \cup \{b\}, E \cup \{\{b, x\}\})$ denote the resulting tree where we assume that G' is not a $(2, n_1, n_2)$-barbell graph. Further, let d_v and d'_v denote the degree of vertex v in G or G', respectively. Similarly, $\mathcal{N}_i(v)$ and $\mathcal{N}'_i(v)$ denote the set of vertices in distance i of v, while $N_i(v) = |\mathcal{N}_i(v)|$ and $N'_i(v) = |\mathcal{N}'_i(v)|$ are the respective cardinalities of these sets.

We have $w'_2 = w_2 + 2d_x + 2$ and $p'_i = p_i + 2N_{i-1}(x)$. Therefore, it is sufficient to show

$$\left(\sum_{v \in V'} N'_2(v) d'_v (2d'_v - 3)\right) - \left(\sum_{v \in V} N_2(v) d_v (2d_v - 3)\right) - 2d_x - 2N_2(x) + p'_5 - p_5 \geq 2 \ .$$

All vertices having distance to b greater than 2 contribute the same value to both sums.

Hence, we obtain

$$
\begin{aligned}
&\left(\sum_{v \in V'} N'_2(v) d'_v (2d'_v - 3)\right) - \left(\sum_{v \in V} N_2(v) d_v (2d_v - 3)\right) \\
=\ & 2{d'_b}^2 N'_2(b) - 3d'_b N'_2(b) + 2{d'_x}^2 N'_2(x) - 3d'_x N'_2(x) - 2d_x^2 N_2(x) + 3d_x N_2(x) \\
& + \left(\sum_{v \in \mathcal{N}'_2(b)} 2{d'_v}^2 N'_2(v) - 3d'_v N'_2(v) - 2d_v^2 N_2(v) + 3d_v N_2(v)\right) \\
=\ & 2d_x - 3d_x + \left[2(d_x + 1)^2 N_2(x) - 3(d_x + 1)N_2(x) - 2d_x^2 N_2(x) + 3d_x N_2(x)\right] \\
& + \left(\sum_{v \in \mathcal{N}_1(x)} 2d_v^2 (N_2(v) + 1) - 3d_v (N_2(v) + 1) - 2d_v^2 N_2(v) + 3d_v N_2(v)\right) \\
=\ & -d_x + 4d_x N_2(x) - N_2(x) + \left(\sum_{v \in \mathcal{N}_1(x)} 2d_v^2 - 3d_v\right) \ .
\end{aligned}
$$

Thus, it is sufficient to show

$$
\begin{aligned}
0 \ \leq \ & -d_x + 4d_x N_2(x) - N_2(x) + \left(\sum_{v \in \mathcal{N}_1(x)} 2d_v^2 - 3d_v \right) \\
& -2d_x - 2N_2(x) + p_5' - p_5 - 2 \\
= \ & 4d_x N_2(x) + \left(\sum_{v \in \mathcal{N}_1(x)} 2d_v^2 - 3d_v + 1 \right) - 4d_x - 3N_2(x) + p_5' - p_5 - 2 \\
= \ & (4d_x - 3) \left[N_2(x) - 1 \right] + \left(\sum_{v \in \mathcal{N}_1(x)} 2d_v^2 - 3d_v + 1 \right) + p_5' - p_5 - 5 \ .
\end{aligned}
$$

Since $\operatorname{diam}(G) \geq 4$, G contains a path with 4 edges as a subgraph. Let c denote the center vertex of this path.

Case 1:

$N_4(x) \neq 0$. This implies $p_5' - p_5 = 2N_4(x) \geq 2$. Since we have $0 \leq (4d_x - 3)(N_2(x) - 1)$, it is sufficient to show $\left(\sum_{v \in \mathcal{N}_1(x)} 2d_v^2 - 3d_v + 1 \right) \geq 3$. Since G has a diameter of at least 4, there must be a neighbor y of x with $d_y \geq 2$. This vertex yields at least $2 \cdot 2^2 - 3 \cdot 2 + 1 = 3$ (all other terms are nonnegative).

Case 2:

$N_4(x) = 0$. Then we have $p_5' = p_5$, $\operatorname{diam}(G) = \operatorname{diam}(G') \leq 6$, and the distance between b and c is at most 2, i.e., $x = c$ or $x \in \mathcal{N}_1(c)$. Hence, $d_x \geq 2$ or $N_2(x) \geq 2$. Now we show

$$
(4d_x - 3)(N_2(x) - 1) + \left(\sum_{v \in \mathcal{N}_1(x)} 2d_v^2 - 3d_v + 1 \right) - 5 \geq 0 \ .
$$

Case 2.1:

$d_x \geq 2$ and $N_2(x) \geq 2$. We obtain

$$
(4d_x - 3)(N_2(x) - 1) - 5 \geq (4 \cdot 2 - 3)(2 - 1) - 5 \geq 0 \ .
$$

Case 2.2:

$d_x = 1$ and $N_2(x) \geq 2$. Then the only neighbor of x must have degree $N_2(x) + 1$. Hence, we have

$$
\begin{aligned}
& (4d_x - 3) \left[N_2(x) - 1 \right] + \left(\sum_{v \in \mathcal{N}_1(x)} 2d_v^2 - 3d_v + 1 \right) - 5 \\
= \ & N_2(x) - 1 + \left(\sum_{v \in \mathcal{N}_1(x)} 2d_v^2 - 3d_v + 1 \right) - 5 \\
= \ & N_2(x) + \left(2[N_2(x) + 1]^2 - 3[N_2(x) + 1] + 1 \right) - 6 \\
= \ & 2N_2(x)^2 + 2N_2(x) - 6 \ > \ 0 \ .
\end{aligned}
$$

Case 2.3:

$d_x \geq 2$ and $N_2(x) = 1$. Then, since $N_4(x) = 0$, the diameter of G is 4, and therefore G' is a $(2, n_1, n_2)$-barbell graph for properly chosen $n_1, n_2 \in \mathbb{N} \setminus \{0\}$. This contradicts the assumption that G' is not such a graph. Thus, this case cannot occur.

Since for every tree having diameter at most 3 and for all $(2, n_1, n_2)$-barbell graphs the inequality $w_0 w_5 \geq w_1 w_4$ is valid as well, it holds for all trees. This finishes the proof of Theorem 5.5.

5.3.4 A Conjecture for Trees

The justification of the inequalities $w_1 w_2 \leq w_0 w_3$ and $w_1 w_4 \leq w_0 w_5$ for trees leads to the following more general conjecture.

Conjecture 5.1
For all trees, the inequality $w_1 \cdot w_k \leq w_0 \cdot w_{k+1}$ is valid for all $k \in \mathbb{N}$, or equivalently, $\overline{d} \cdot w_k \leq w_{k+1}$ for all non-empty trees.

Then, in contrast to general graphs, trees would also observe the inequality for all odd (not only even) indices on the greater side. This case of an *odd* index on the greater side is equivalent to the following statement about averages:

$$\frac{1}{n} \sum_{x \in V} w_k(x)^2 \leq \frac{1}{m} \sum_{\{x,y\} \in E} w_k(x) \cdot w_k(y) \ .$$

5.4 Subdivision Graphs

For a given graph $G = (V, E)$, the *subdivision graph* S_G is obtained from G by introducing for each edge $e \in E$ a new vertex v_e that splits the old edge $e = \{v_1, v_2\}$ and replaces it by two new edges $e_1 = \{v_1, v_e\}$ and $e_2 = \{v_e, v_2\}$.

5.4.1 Walks of Length 2 and 3

Ilić and Stevanović [IS09] proved the following theorem for subdivision graphs.

Theorem 5.6 (Ilić and Stevanović)
For every undirected graph G on n vertices and m edges, the corresponding subdivision graph S_G obeys the inequality

$$\frac{M_1(S_G)}{n+m} \leq \frac{M_2(S_G)}{2m} \ .$$

If we translate this to walk numbers, this statement corresponds to the following.

Corollary 5.2

For every undirected graph G, the corresponding subdivision graph S_G obeys the inequality

$$w_1(S_G) \cdot w_2(S_G) \leq w_0(S_G) \cdot w_3(S_G) \ .$$

Proof We have

$$
\begin{aligned}
w_0(S_G) &= n + m = w_0 + w_1/2 &&\text{(We have } m \text{ new vertices.)}\\
w_1(S_G) &= 4m = 2w_1 &&\text{(Every edge is split into two parts.)}\\
w_2(S_G) &= \left(\sum_{v \in V} d_v^2\right) + m \cdot 2^2 = w_2 + 4m = w_2 + 2w_1 \\
w_3(S_G) &= 2 \sum_{\{x,y\} \in E(S_G), x \in V} 2d_x = 4 \sum_{x \in V} d_x^2 = 4w_2
\end{aligned}
$$

Then the inequality corresponds to

$$
\begin{aligned}
(2w_1)(w_2 + 2w_1) &\leq (w_0 + w_1/2)(4w_2)\\
2w_1w_2 + 4w_1^2 &\leq 2w_1w_2 + 4w_0w_2\\
w_1^2 &\leq w_0w_2 \ .
\end{aligned}
$$

The inequality in the last line is valid by Theorem 4.1.

5.4.2 Walks of Length 4 and 5

Now, it would be interesting to know whether a similar inequality holds for longer walks in subdivision graphs, e.g., for the number of 4-step and 5-step walks (see also [Täu15b]).

Theorem 5.7

For every undirected graph G, the corresponding subdivision graph S_G obeys the inequality

$$w_1(S_G) \cdot w_4(S_G) \leq w_0(S_G) \cdot w_5(S_G) \ .$$

Proof The calculation starts by applying the formula $w_4(S_G) = \sum_{v \in V(S_G)} w_2(v)^2$. We need to distinguish two kinds of vertices: old vertices (corresponding to the vertices in the original graph G) and new vertices (corresponding to the edges of G). For each old vertex $v \in V$, the number of 2-step walks in S_G is $w_2^{S_G}(v) = 2d_v$. For every new vertex v_e corresponding to an edge $e = \{x, y\}$ in G, the number of 2-step walks in S_G is $w_2^{S_G}(v_e) = d_x + d_y$. Hence we obtain the following walk numbers in terms of the degrees and walks of the original graph G:

$$
\begin{aligned}
w_4(S_G) &= \sum_{v \in V}(2d_v)^2 + \sum_{\{x,y\} \in E}(d_x + d_y)^2 = 4w_2 + \sum_{\{x,y\} \in E}(d_x + d_y)^2 \\
w_5(S_G) &= 2\sum_{\{x,y\} \in E} 2d_x \cdot (d_x + d_y) + 2d_y \cdot (d_x + d_y) = 4\sum_{\{x,y\} \in E}(d_x + d_y)^2
\end{aligned}
$$

We obtain the following equivalent inequalities:

$$
\begin{aligned}
(2w_1)\left(4w_2 + \sum_{\{x,y\} \in E}(d_x + d_y)^2\right) &\leq (w_0 + w_1/2)\left(4\sum_{\{x,y\} \in E}(d_x + d_y)^2\right) \\
8w_1 w_2 &\leq 4w_0 \sum_{\{x,y\} \in E}(d_x + d_y)^2 \\
2\frac{w_1}{w_0} &\leq \frac{1}{w_2}\sum_{\{x,y\} \in E}(d_x + d_y)^2
\end{aligned}
$$

The last inequality follows from Theorem 4.36.

Theorem 5.8
For every graph G, the corresponding subdivision graph S_G obeys the inequality

$$
w_2(S_G) \cdot w_3(S_G) \leq w_0(S_G) \cdot w_5(S_G) .
$$

Proof We have

$$
\begin{aligned}
(w_2 + 2w_1)(4w_2) &\leq (w_0 + w_1/2)\left(4\sum_{\{x,y\} \in E}(d_x + d_y)^2\right) \\
w_2^2 + 2w_1 w_2 &\leq (w_0 + w_1/2)\left(\sum_{\{x,y\} \in E}(d_x + d_y)^2\right)
\end{aligned}
$$

The inequality follows from Corollary 4.21 and Theorem 4.36 after observing that $w_1/2 = m$.

DIRECTED GRAPHS / NONSYMMETRIC MATRICES

III

Chapter 6

Walks and Alternating Walks in Directed Graphs

A large part of mathematics which becomes useful developed with absolutely no desire to be useful, and in a situation where nobody could possibly know in what area it would become useful; and there were no general indications that it ever would be so.

John von Neumann

6.1 Chebyshev's Sum Inequality

In order to obtain inequalities for the number of walks in directed graphs and for entry sums in nonsymmetric matrices, it is sometimes possible to apply Chebyshev's sum inequality (see Theorems 1.6 and 1.7). In those cases we are able to obtain statements by elementary proofs without using any eigenvalues.

Theorem 6.1
For any matrix A such that the column sums of A^k and the row sums of A^ℓ (i.e., $c^{[k]}$ and $r^{[\ell]}$) are similarly ordered, we have

$$\text{sum}\left(A^k\right) \cdot \text{sum}\left(A^\ell\right) \le n \cdot \text{sum}\left(A^{k+\ell}\right) .$$

The inequality is reversed if $c^{[k]}$ and $r^{[\ell]}$ are conversely ordered.

Proof   The inequality is a direct consequence of Corollary 1.1 and Chebyshev's inequality (see Theorem 1.6):

$$\operatorname{sum}\left(A^k\right) \cdot \operatorname{sum}\left(A^\ell\right) = \left(\sum_{i=1}^n c_i^{[k]}\right)\left(\sum_{i=1}^n r_i^{[\ell]}\right) \le n \sum_{i=1}^n c_i^{[k]} r_i^{[\ell]} = n \cdot \operatorname{sum}\left(A^{k+\ell}\right) \ .$$

For the special case of adjacency matrices, this translates to the following statement about the number of walks in digraphs.

Corollary 6.1
For every directed graph $G = (V, E)$ where the vectors of walk numbers $e_k(v)$ and $s_\ell(v)$, $v \in V$, are similarly ordered, we have

$$w_k \cdot w_\ell \le n \cdot w_{k+\ell} \ .$$

Obviously, this inequality is applicable to undirected graphs if $w_k(v_i)$ and $w_\ell(v_i)$, $i \in [n]$, are similarly ordered sequences (here, we have $w_k(v_i) = s_k(v_i) = e_k(v_i)$ for all $i, k \in \mathbb{N}$). In particular, this is interesting if $k + \ell$ is an odd number.

According to Chebyshev's sum inequality (see Theorem 1.6), the inequality is inverted if $e_k(v_i)$ and $s_\ell(v_i)$ are conversely ordered. For instance, this would be applicable for $k = \ell = 1$ if for each vertex either the in-degree or the out-degree is equal to 1 and the other one is greater or equal to 1. Another example would be the class of graphs where all vertices have the same sum of the in-degree and the out-degree (that is, the same total degree).

From Theorem 6.1, we obtain a special case if the row sums and the column sums of a matrix are similarly ordered. This happens, for example, in the case of sum-symmetric matrices, i.e., if $r_i(A) = c_i(A)$ for all $i \in [n]$.

Corollary 6.2
For any sum-symmetric matrix A, we have

$$\operatorname{sum}(A)^2 \le n \cdot \operatorname{sum}(A^2) \ .$$

Note that this corollary also follows from an inequality similar to Lemma 1.4. Since $f(x) = x^2$ is convex in $\mathbb{R}$, the assumption $a_i \ge 0$ in Lemma 1.4 is not necessary for $p = 2$.

$$\operatorname{sum}(A)^2 = \left(\sum_{i=1}^n r_i\right)^2 \le n \sum_{i=1}^n r_i^2 = n \sum_{i=1}^n r_i c_i = n \operatorname{sum}(A^2) \ .$$

Now we apply the method to directed graphs. If there is a vertex ordering which is monotonically increasing with respect to the in- and out-degrees, then the graph obeys the inequality $n w_2 \ge w_1^2$. For instance, this is true if the in-degree of each vertex equals its out-degree.

Corollary 6.3
For every Eulerian directed graph $(d_{in}(v) = d_{out}(v)$ for all $v \in V)$, we have

$$w_1^2 \leq n \cdot w_2 \ .$$

6.2 Relaxed Inequalities Using Geometric Means

Recall that $\mathfrak{A}$ and $\mathfrak{G}$ denote the arithmetic and geometric means of the given terms.

Theorem 6.2
For every matrix with nonnegative column and row sums, we have

$$\left(\underset{i \in [n]}{\mathfrak{G}}\, c_i^{[k]}\right)\left(\underset{i \in [n]}{\mathfrak{G}}\, r_i^{[\ell]}\right) \leq \underset{i \in [n]}{\mathfrak{A}}\, c_i^{[k+\ell]} = \underset{i \in [n]}{\mathfrak{A}}\, r_i^{[k+\ell]} = \frac{\mathrm{sum}(A^{k+\ell})}{n} \ .$$

Proof By Corollary 1.1 and by the inequality of arithmetic and geometric means, we have

$$\sqrt[n]{\prod_{i\in[n]} c_i^{[k]}} \cdot \sqrt[n]{\prod_{i\in[n]} r_i^{[\ell]}} = \sqrt[n]{\prod_{i\in[n]} c_i^{[k]}\cdot r_i^{[\ell]}} \leq \frac{1}{n}\sum_{i\in[n]} c_i^{[k]}\cdot r_i^{[\ell]} = \frac{1}{n}\sum_{i\in[n]} c_i^{[k+\ell]} = \frac{1}{n}\sum_{i\in[n]} r_i^{[k+\ell]}$$

$$\left(\underset{i \in [n]}{\mathfrak{G}}\, c_i^{[k]}\right)\left(\underset{i \in [n]}{\mathfrak{G}}\, r_i^{[\ell]}\right) = \left(\underset{i \in [n]}{\mathfrak{G}}\, c_i^{[k]}r_i^{[\ell]}\right) \leq \frac{1}{n}\,\mathrm{sum}\left(A^{k+\ell}\right) = \underset{i \in [n]}{\mathfrak{A}}\, c_i^{[k+\ell]} = \underset{i \in [n]}{\mathfrak{A}}\, r_i^{[k+\ell]}$$

Corollary 6.4
For every directed graph, we have

$$\left(\underset{v \in V}{\mathfrak{G}}\, e_k(v)\right)\left(\underset{v \in V}{\mathfrak{G}}\, s_\ell(v)\right) \leq \underset{v \in V}{\mathfrak{A}}\, e_{k+\ell}(v) = \underset{v \in V}{\mathfrak{A}}\, s_{k+\ell}(v) = \frac{w_{k+\ell}}{n} \ .$$

6.3 Alternating Matrix Powers and Alternating Walks

6.3.1 Definitions and Interpretations

Within this section, we mainly deal with expressions that are derived from matrix products of the form $[A]A^*AA^* \ldots A[A^*]$.[1] We call these terms *alternating powers* of a matrix A. In contrast to the conventional matrix power, this kind of alternating matrix power can also be applied to rectangular matrices. Alternating powers of even order (i.e., $(AA^*)^k$ and $(A^*A)^k$ for $k \in \mathbb{N}$) represent positive-semidefinite Hermitian matrices.

For real matrices, the form corresponds to $[A]A^T AA^T \ldots A[A^T]$ (and the alternating powers of even order are symmetric matrices). This implies an interesting analogon

---

[1] Here, the square brackets shall indicate that the sequence could start and end optionally with A or A^T.

for walks in directed graphs. The (i, j)-entry of an alternating power of the adjacency matrix A of a directed graph G corresponds to the number of walks from v_i to v_j that use the directed edges of E in alternating forward and backward direction for each step, corresponding to the pattern of A and A^T. We call this an *alternating walk* in a directed graph.

The alternating powers can be specified by the starting term and the number of factors (i.e., the starting direction and the number of steps). Similar to the number of conventional walks, let $\mathrm{alt}_\ell^{\rightarrow}$ denote the number of alternating ℓ-step walks that start traversing an edge in forward direction. Accordingly, $\mathrm{alt}_\ell^{\leftarrow}$ denotes the number of alternating ℓ-step walks that start traversing an edge in backward direction. For specifying certain start and end vertices, we use a similar notation as for the conventional walks.

In a directed graph with adjacency matrix A, we have

$$\mathrm{alt}_2^{\rightarrow} = \mathrm{sum}(AA^T) = \sum_{v \in V} d_{\mathrm{in}}(v)^2 \qquad \text{and} \qquad \mathrm{alt}_2^{\leftarrow} = \mathrm{sum}(A^T A) = \sum_{v \in V} d_{\mathrm{out}}(v)^2 \ .$$

For any complex matrix A with row sums vector r and column sums vector c, we have

$$\mathrm{sum}(AA^*) = \sum_{i \in [n]} c_i \overline{c_i} = \sum_{i \in [n]} |c_i|^2 = \langle c, c \rangle$$

and

$$\mathrm{sum}(A^* A) = \sum_{i \in [n]} r_i \overline{r_i} = \sum_{i \in [n]} |r_i|^2 = \langle r, r \rangle \ .$$

For a real matrix A, this corresponds to

$$\mathrm{sum}\left(AA^T\right) = \sum_{i=1}^n c_i^2 \qquad \text{and} \qquad \mathrm{sum}\left(A^T A\right) = \sum_{i=1}^n r_i^2 \ .$$

By applying Corollary 1.1 and the Cauchy-Schwarz inequality, we have

$$\left(\mathrm{sum}\left(A^2\right)\right)^2 = \left(\sum_{i=1}^n c_i r_i\right)^2 \le \left(\sum_{i=1}^n c_i \overline{c_i}\right)\left(\sum_{i=1}^n r_i \overline{r_i}\right) = \mathrm{sum}\left(AA^*\right) \cdot \mathrm{sum}\left(A^* A\right)$$

For real matrices, this means

$$\left(\mathrm{sum}\left(A^2\right)\right)^2 = \left(\sum_{i=1}^n c_i r_i\right)^2 \le \left(\sum_{i=1}^n c_i^2\right)\left(\sum_{i=1}^n r_i^2\right) = \mathrm{sum}\left(AA^T\right) \cdot \mathrm{sum}\left(A^T A\right)$$

For directed graphs, this translates to

$$w_2^2 = \left[\sum_{v \in V} d_{\mathrm{in}}(v) d_{\mathrm{out}}(v)\right]^2 \le \left[\sum_{v \in V} d_{\mathrm{in}}(v)^2\right]\left[\sum_{v \in V} d_{\mathrm{out}}(v)^2\right] = \mathrm{alt}_2^{\rightarrow} \cdot \mathrm{alt}_2^{\leftarrow}.$$

Undirected Bipartite Graphs.

Instead of considering alternating walks in an arbitrary directed graph, alternating matrix powers can also be interpreted as walks in an *undirected bipartite graph*. The

$m \times n$ matrix A is regarded as some kind of reduced adjacency matrix (Brualdi [Bru11] calls it a *biadjacency matrix*) between m red vertices and n blue vertices (see Banks, Harcharras, Neuwirth and Ricard [BHNR03] and Pate [Pat12]). Then, $\text{sum}(AA^T A)$ is the number of walks of length 3 starting at a red vertex. Similarly, $\text{sum}(A^T AA^T)$ is the number of walks of length 3 starting at a blue vertex. Thus, $w_3 = \text{sum}(AA^T A) + \text{sum}(A^T AA^T)$. Since $(A^T AA^T)^T = AA^T A$ (i.e., the number of walks of length 3 starting at red vertices equals the number of walks of length 3 starting at blue vertices), we have $w_3 = 2 \cdot \text{sum}(AA^T A)$. This is also clear from the fact that each such walk starts at a red vertex and ends at a blue vertex, i.e., it could be reversed for counting. This is not true for an even number of steps. More generally, we see that $\text{sum}((AA^T)^k A) = \text{sum}((A^T A)^k A^T)$ since $((AA^T)^k A)^T = (A^T A)^k A^T$ while $\text{sum}((AA^T)^k) \neq \text{sum}((A^T A)^k)$ and $((AA^T)^k)^T = (AA^T)^k \neq (A^T A)^k$.

6.3.2 A Simple Generalization

The following is a generalization of Theorem 4.2 (Dress and Gutman [DG03b]), Theorem 4.5 (Marcus and Newman [MN62]), and our Theorem 4.8.

Theorem 6.3
For all complex matrices $A, B \in \mathbb{C}^{n \times n}$ and complex vectors $x, y \in \mathbb{C}^n$, we have

$$|\text{sum}_{x,y}(A^* B)|^2 = \text{sum}_{x,y}(A^* B) \cdot \text{sum}_{y,x}(B^* A) \leq \text{sum}_x(A^* A) \cdot \text{sum}_y(B^* B) \ .$$

Proof According to the definition, the statement is translated to

$$|x^* A^* By|^2 = x^* A^* By \cdot y^* B^* Ax \leq x^* A^* Ax \cdot y^* B^* By \ .$$

Setting $a := Ax$ and $b := By$, this is equivalent to

$$|a^* b|^2 = a^* bb^* a \leq a^* ab^* b$$
$$|\langle a, b \rangle|^2 = \langle a, b \rangle \langle b, a \rangle \leq \langle a, a \rangle \langle b, b \rangle \ ,$$

which is valid by the Cauchy-Schwarz inequality (see Theorem 1.2).

Corollary 6.5
For every directed graph $G = (V, E)$, vertex subsets $X, Y \subseteq V$, and all $a, b \in \mathbb{N}$, the following inequalities hold.
If a and b are both even or both odd numbers, then we have

$$\text{alt}^{\rightarrow}_{a+b}(X, Y)^2 \leq \text{alt}^{\rightarrow}_{2a}(X, X) \cdot \text{alt}^{\rightarrow}_{2b}(Y, Y)$$
$$\text{alt}^{\leftarrow}_{a+b}(X, Y)^2 \leq \text{alt}^{\leftarrow}_{2a}(X, X) \cdot \text{alt}^{\leftarrow}_{2b}(Y, Y) \ .$$

If one of the numbers a and b is even and the other one is odd, then we have

$$\text{alt}^{\rightarrow}_{a+b}(X, Y)^2 \leq \text{alt}^{\rightarrow}_{2a}(X, X) \cdot \text{alt}^{\leftarrow}_{2b}(Y, Y)$$
$$\text{alt}^{\leftarrow}_{a+b}(X, Y)^2 \leq \text{alt}^{\leftarrow}_{2a}(X, X) \cdot \text{alt}^{\rightarrow}_{2b}(Y, Y) \ .$$

Proof   Apply Theorem 6.3 to the characteristic vectors ψ_X and ψ_Y of the vertex subsets $X, Y \subseteq V$ and the a-th and b-th alternating powers of the adjacency matrix of G.

We will discuss more general results in a later part of this section.

6.3.3   Related Work

At about the same time when Mulholland and Smith [MS59] published Theorem 4.20, Atkinson, Watterson and Moran [AWM60] obtained the following result. Both of the statements provided a proof for a conjecture of Mandel and Hughes [MH58] in population genetics (see Section 2.8).

Theorem 6.4 (Atkinson et al.)
Let A be a nonnegative $m \times n$ matrix. Then

$$\mathrm{sum}(A)^3 \leq mn \sum_{i=1}^{m} \sum_{j=1}^{n} a_{ij} c_j r_i \ .$$

This is equivalent to

$$\left(\sum_{i=1}^{m} \sum_{j=1}^{n} a_{ij} \right)^3 \leq mn \sum_{i=1}^{m} \sum_{j=1}^{n} a_{ij} \sum_{r=1}^{m} a_{rj} \sum_{s=1}^{n} a_{is}$$

and

$$\mathrm{sum}(A)^3 \leq mn \cdot \mathrm{sum}(AA^T A) = mn \cdot \mathrm{sum}(A^T AA^T) \ .$$

Theorem 6.4 was used by Atkinson et al. to show

$$\left(\sum_{i=1}^{m} \sum_{j=1}^{n} a_{ij} p_i q_j \right)^3 \leq \sum_{i=1}^{m} \sum_{j=1}^{n} a_{ij} p_i q_j \sum_{r=1}^{m} a_{rj} p_r \sum_{s=1}^{n} a_{is} q_s$$

for $a_{ij} > 0$ and nonnegative numbers p_i, q_j with $\sum_{i=1}^{m} p_i = \sum_{j=1}^{n} q_j = 1$. For symmetric matrices ($m = n$, $a_{ij} = a_{ji}$) and equal weights ($p_i = q_i$), this is equivalent to the conjecture of Mandel and Hughes.

Kingman [Kin61b] gave a more direct proof of Atkinson et al.'s result using the convexity of x^k. Kingman also showed that it is a special case of a class of inequalities involving sets of nonnegative numbers depending on several indices, and 'partial averages' over subsets of those indices. In another paper, Kingman [Kin67] remarks that such a partial average is just a particular example of a conditional expectation, and a conditional expectation is a special sort of Radon-Nikodym derivative. Then he showed an inequality for such derivatives that generalizes the former results.

Banks, Harcharras, Neuwirth and Ricard [BHNR03] generalized Theorem 6.4 in the following way.

Theorem 6.5 (Banks et al.)
Let A be an $m \times n$-matrix with nonnegative real entries. Then for every integer $k \geq 1$, the following inequalities hold:

$$\text{sum}(A)^{2k} \leq m^{k-1}n^k \,\text{sum}\left((AA^T)^k\right)$$
$$\text{sum}(A)^{2k+1} \leq m^k n^k \,\text{sum}\left((AA^T)^k A\right) \ .$$

This, in turn, is a special case of a recent result of Pate [Pat12] that generalizes Theorems 4.20 and 4.22 (the inequality of Mulholland and Smith; Blakley and Roy) to nonsymmetric rectangular matrices and two possibly different scaling vectors.

Lemma 6.1 (Pate)
Suppose that A is an $n \times n$ positive-semidefinite Hermitian matrix with complex entries. If x is a unit vector in $\mathbb{C}^n$, and $\alpha > 1$, then

$$\langle A^\alpha x, x \rangle \geq \langle Ax, x \rangle^\alpha \ ,$$

with equality if and only if x is an eigenvector of A.

This lemma was used to show the following generalization of Theorems 4.20 and 4.22 for even powers.

Theorem 6.6 (Pate)
Suppose $m, n \in \mathbb{N}$, and A is a complex $m \times n$ matrix. If $u \in \mathbb{C}^n$, $v \in \mathbb{C}^m$, and $\|u\| = \|v\| = 1$, then

$$\langle (A^*A)^\alpha u, u \rangle \geq |\langle Au, v \rangle|^{2\alpha} \ , \qquad \forall \alpha > 1 \ ,$$

with equality if and only if $Au = 0$, or there exists a nonzero $\zeta \in \mathbb{C}$ such that $Au = \zeta v$ and $A^*v = \bar{\zeta}u$.

Then, Pate showed a quite surprising result that generalizes Theorems 4.20 and 4.22 for odd powers. Note here, that the conceivable inequality

$$\langle (A^TA)^k u, v \rangle \geq (\langle Au, v \rangle)^{2k}$$

for even powers is *not* valid in general for nonnegative square matrices $A \in \mathbb{R}^{n \times n}$, nonnegative unit vectors $u \in \mathbb{R}^n$ and $v \in \mathbb{R}^n$, and $k \in \mathbb{N}$.

Theorem 6.7 (Pate)
For any nonnegative $m \times n$ matrix A and nonnegative unit vectors $u \in \mathbb{R}^n_{\geq 0}$ and $v \in \mathbb{R}^m_{\geq 0}$, $\|u\| = \|v\| = 1$, we have

$$\left\langle \left(AA^T\right)^k Au, v \right\rangle \geq \langle Au, v \rangle^{2k+1} \ .$$

As a special case, Pate deduced again the inequalities of Theorem 6.5. Theorem 6.7 settles a special case of Sidorenko's Conjecture (see Conjecture 9.2). Pate's inequalities were also used in the context of normalized cuts-based clustering, see Peluffo-Ordóñez, Castro-Hoyos, Acosta-Medina and Castellanos-Domínguez [PCAC14]. Alternating walks were also used by Butler [But06a; But06b; But08] in the context of discrepancy.

6.3.4 Transferring Hermitian Matrix Results to General Matrices

In order to deduce a method for deriving inequalities on arbitrary rectangular matrices from the results for symmetric or Hermitian matrices, we propose a remarkably simple procedure. The idea is as follows. From a given $m \times n$-matrix A, we construct an auxiliary block matrix[2]

$$B := \begin{pmatrix} 0 & A \\ A^* & 0 \end{pmatrix} .$$

The two zeros indicate two square submatrices of size $m \times m$ and $n \times n$ that have only zero entries. Notice that B is a Hermitian $(m+n) \times (m+n)$-matrix, that means, we can apply all results for Hermitian matrices to B. Now it is interesting to see what happens to the powers of B. For the first powers, we have

$$B^0 = \begin{pmatrix} I_m & 0 \\ 0 & I_n \end{pmatrix}, \ B^1 = \begin{pmatrix} 0 & A \\ A^* & 0 \end{pmatrix}, \ B^2 = \begin{pmatrix} AA^* & 0 \\ 0 & A^*A \end{pmatrix}, B^3 = \begin{pmatrix} 0 & AA^*A \\ A^*AA^* & 0 \end{pmatrix}.$$

More generally, for $k \in \mathbb{N}$, we have

$$\begin{pmatrix} 0 & A \\ A^* & 0 \end{pmatrix}^{2k} = \begin{pmatrix} (AA^*)^k & 0 \\ 0 & (A^*A)^k \end{pmatrix} \text{ and } \begin{pmatrix} 0 & A \\ A^* & 0 \end{pmatrix}^{2k+1} = \begin{pmatrix} 0 & (AA^*)^kA \\ (A^*A)^kA^* & 0 \end{pmatrix}.$$

Lemma 6.2
For every complex $m \times n$-matrix A, all vectors $u \in \mathbb{C}^m$ and $v \in \mathbb{C}^n$, and all $k \in \mathbb{N}$, we have

$$\begin{pmatrix} u \\ v \end{pmatrix}^* \begin{pmatrix} 0 & A \\ A^* & 0 \end{pmatrix}^{2k+1} \begin{pmatrix} u \\ v \end{pmatrix} = u^*(AA^*)^kAv + v^*(A^*A)^kA^*u$$

$$= 2\Re\left(u^*(AA^*)^kAv\right) \qquad and$$

$$\begin{pmatrix} u \\ v \end{pmatrix}^* \begin{pmatrix} 0 & A \\ A^* & 0 \end{pmatrix}^{2k} \begin{pmatrix} u \\ v \end{pmatrix} = u^*(AA^*)^ku + v^*(A^*A)^kv .$$

---

[2]It turned out that this block matrix has already been used by Jordan and others to transfer eigenvalue results from Hermitian matrices to singular value results for general matrices, see Horn and Johnson [HJ91, p. 135].

Proof   From the preceding observations, we know

$$\begin{pmatrix} u \\ v \end{pmatrix}^* \begin{pmatrix} 0 & A \\ A^* & 0 \end{pmatrix}^{2k+1} \begin{pmatrix} u \\ v \end{pmatrix} = \begin{pmatrix} u \\ v \end{pmatrix}^* \begin{pmatrix} 0 & (AA^*)^k A \\ (A^*A)^k A^* & 0 \end{pmatrix} \begin{pmatrix} u \\ v \end{pmatrix}$$

$$= u^*(AA^*)^k Av + v^*(A^*A)^k A^* u$$

$$= u^*(AA^*)^k Av + \left(u^*(AA^*)^k Av\right)^*$$

$$= 2\Re\left(u^*(AA^*)^k Av\right)$$

and

$$\begin{pmatrix} u \\ v \end{pmatrix}^* \begin{pmatrix} 0 & A \\ A^* & 0 \end{pmatrix}^{2k} \begin{pmatrix} u \\ v \end{pmatrix} = \begin{pmatrix} u \\ v \end{pmatrix}^* \begin{pmatrix} (AA^*)^k & 0 \\ 0 & (A^*A)^k \end{pmatrix} \begin{pmatrix} u \\ v \end{pmatrix}$$

$$= u^*(AA^*)^k u + v^*(A^*A)^k v \ .$$

6.3.5   *Generalizations for Nonnegative Matrices*

Now, we show an alternative proof of Theorem 6.7. This proof is not only very simple, it also shows that Theorems 4.20 and 4.22 are equivalent to Theorem 6.7 in some sense. As an important consequence, it explains why the invalid inequality for even powers in Pate's paper is not the 'proper translation'. Our method of proof provides a corresponding valid form for this case.

Theorem 6.8
  For any nonnegative $m \times n$ matrix A, nonnegative vectors $u \in \mathbb{R}^m_{\geq 0}$ and $v \in \mathbb{R}^n_{\geq 0}$, and all $k \in \mathbb{N}$, we have

$$\left(u^T Av\right)^{2k+1} \leq \left(\frac{\|u\|^2 + \|v\|^2}{2}\right)^{2k} u^T \left(AA^T\right)^k Av \qquad and$$

$$\left(u^T Av\right)^{2k} \leq \left(\frac{\|u\|^2 + \|v\|^2}{2}\right)^{2k-1} \frac{u^T \left(AA^T\right)^k u + v^T \left(A^T A\right)^k v}{2} \ .$$

Proof   If A, u, and v are nonnegative, then we can apply Theorem 4.20 to the symmetric matrix $B := \begin{pmatrix} 0 & A \\ A^T & 0 \end{pmatrix}$ and to the vector s that is composed of the two vectors u and v. We conclude $s^T B^\ell s (s^T s)^{\ell-1} \geq (s^T Bs)^\ell$. Now we apply Lemma 6.2 to B^ℓ. For *odd* exponents $\ell = 2k + 1$, we obtain

$$s^T B^{2k+1} s (s^T s)^{2k} \geq (s^T Bs)^{2k+1}$$

$$\left(2u^T (AA^T)^k Av\right) \left(\|u\|^2 + \|v\|^2\right)^{2k} \geq \left(2u^T Av\right)^{2k+1} \ .$$

For *even* exponents $\ell = 2k$, we have

$$s^T B^{2k} s (s^T s)^{2k-1} \geq (s^T Bs)^{2k}$$

$$\left(u^T (AA^T)^k u + v^T (A^T A)^k v\right) \left(\|u\|^2 + \|v\|^2\right)^{2k-1} \geq \left(2u^T Av\right)^{2k} \ .$$

Corollary 6.6
For any nonnegative $m \times n$ matrix A, subsets $U \subseteq [m]$ and $V \subseteq [n]$, and all $k \in \mathbb{N}$, we have

$$\mathrm{sum}\,(A[U,V])^{2k+1} \leq \left(\frac{|U|+|V|}{2}\right)^{2k} \mathrm{sum}\left(((AA^T)^k A)[U,V]\right) \qquad and$$

$$\mathrm{sum}\,(A[U,V])^{2k} \leq \left(\frac{|U|+|V|}{2}\right)^{2k-1} \frac{\mathrm{sum}\left(((AA^T)^k)[U]\right) + \mathrm{sum}\left(((A^T A)^k)[V]\right)}{2}.$$

The resulting consequences for the corresponding numbers of alternating walks in directed graphs are obvious. For the special case $U = [m]$ and $V = [n]$, we obtain the following.

Corollary 6.7
For any nonnegative $m \times n$ matrix A and all $k \in \mathbb{N}$, we have

$$\mathrm{sum}\,(A)^{2k+1} \leq \left(\frac{m+n}{2}\right)^{2k} \mathrm{sum}((AA^T)^k A) \qquad and$$

$$\mathrm{sum}\,(A)^{2k} \leq \left(\frac{m+n}{2}\right)^{2k-1} \frac{\mathrm{sum}((AA^T)^k) + \mathrm{sum}((A^T A)^k)}{2}.$$

If u and v are nonnegative *unit* vectors, then the inequalities in Theorem 6.8 reduce to the following.

Theorem 6.9
For any nonnegative $m \times n$ matrix A, nonnegative unit vectors $u \in \mathbb{R}^m_{\geq 0}$, $v \in \mathbb{R}^n_{\geq 0}$, $\|u\| = \|v\| = 1$, and all $k \in \mathbb{N}$, we have

$$\left(u^T A v\right)^{2k+1} \leq u^T \left(AA^T\right)^k A v \qquad and$$

$$\left(u^T A v\right)^{2k} \leq \frac{u^T \left(AA^T\right)^k u + v^T \left(A^T A\right)^k v}{2}.$$

The first part corresponds to Theorem 6.7. The second part explains the case of even exponents. The special cases of Theorems 6.8 and 6.9 where $u = v$ and A is symmetric correspond to the Theorems 4.20 and 4.22 by Mulholland and Smith [MS59] and Blakley and Roy [BR65].

If we apply Theorem 6.9 to normalized vectors, we obtain the following form.

Corollary 6.8
For any nonnegative $m \times n$ matrix A, nonnegative vectors $u \in \mathbb{R}_{\geq 0}^m$ and $v \in \mathbb{R}_{\geq 0}^n$, and all $k \in \mathbb{N}$, we have

$$\left(u^T A v\right)^{2k+1} \leq \left(u^T u \cdot v^T v\right)^k u^T \left(A A^T\right)^k A v \qquad \text{and}$$

$$\left(u^T A v\right)^{2k} \leq \frac{1}{2} \left(u^T u \cdot v^T v\right)^k \left(\frac{u^T \left(A A^T\right)^k u}{u^T u} + \frac{v^T \left(A^T A\right)^k v}{v^T v}\right) .$$

Corollary 6.9
For any nonnegative $m \times n$ matrix A, subsets $U \subseteq [m]$ and $V \subseteq [n]$, and all $k \in \mathbb{N}$, we have

$$\operatorname{sum}\left(A[U, V]\right)^{2k+1} \leq \left(|U| \cdot |V|\right)^k \operatorname{sum}\left(\left((A A^T)^k A\right)[U, V]\right) \qquad \text{and}$$

$$\operatorname{sum}\left(A[U, V]\right)^{2k} \leq \frac{1}{2}\left(|U| \cdot |V|\right)^k \left(\frac{\operatorname{sum}\left(\left((A A^T)^k\right)[U]\right)}{|U|} + \frac{\operatorname{sum}\left(\left((A^T A)^k\right)[V]\right)}{|V|}\right) .$$

Again, there are obvious consequences for the corresponding numbers of alternating walks in directed graphs. For the special case $U = [m]$ and $V = [n]$, we obtain the following.

Corollary 6.10
For any nonnegative $m \times n$ matrix A and all $k \in \mathbb{N}$, we have

$$\operatorname{sum}(A)^{2k+1} \leq (mn)^k \operatorname{sum}((A A^T)^k A) \qquad \text{and}$$

$$\operatorname{sum}(A)^{2k} \leq \frac{1}{2}(mn)^k \left(\frac{\operatorname{sum}((A A^T)^k)}{m} + \frac{\operatorname{sum}((A^T A)^k)}{n}\right) .$$

Notice that these inequalities can be interpreted as some kind of matrix densities result. To this end, we would write it in a slightly different way:

$$\left(\frac{\operatorname{sum}(A)}{\sqrt{mn}}\right)^{2k+1} \leq \frac{\operatorname{sum}((A A^T)^k A)}{\sqrt{mn}} \qquad \text{and}$$

$$\left(\frac{\operatorname{sum}(A)}{\sqrt{mn}}\right)^{2k} \leq \frac{1}{2}\left(\frac{\operatorname{sum}((A A^T)^k)}{m} + \frac{\operatorname{sum}((A^T A)^k)}{n}\right) .$$

Something similar can be applied to Corollary 6.7, using the arithmetic mean $(m + n)/2$ instead of the geometric mean $\sqrt{mn}$.

6.3.5.1 *Implications for Hermitian Matrices*

Now, we obtain some results for *Hermitian* (in particular, for *nonnegative symmetric*) matrices that are implied by Theorem 6.8. Assume that A is a Hermitian $n \times n$-matrix, $x, y \in \mathbb{C}^n$ are vectors, and $a, b, p \in \mathbb{N}$ are integers such that A^p is a nonnegative symmetric matrix and $u := A^a x$ and $v := A^b y$ are nonnegative vectors ($u, v \in \mathbb{R}_{\geq 0}^n$). This applies, in particular, if A is a nonnegative symmetric matrix and x, y are nonnegative vectors. Since u and v are nonnegative (i.e., real) vectors, we have $(A^a x)^T = (A^a x)^* = x^*(A^*)^a$ and $(A^b y)^T = (A^b y)^* = y^*(A^*)^b$. Since A is Hermitian, we have $A^* = A$ and Theorem 6.8 implies

$$\left(x^* A^a A^p A^b y\right)^{2k+1} \leq \left(\frac{x^* A^a A^a x + y^* A^b A^b y}{2}\right)^{2k} x^* A^a A^{p(2k+1)} A^b y \qquad \text{and}$$

$$\left(x^* A^a A^p A^b y\right)^{2k} \leq \left(\frac{x^* A^a A^a x + y^* A^b A^b y}{2}\right)^{2k-1} \frac{x^* A^a A^{p(2k)} A^a x + y^* A^b A^{p(2k)} A^b y}{2}.$$

Corollary 6.11

For any Hermitian $n \times n$-matrix A, vectors $x, y \in \mathbb{C}^n$, and all $a, b, k, p \in \mathbb{N}$ such that A^p, $A^a x$, and $A^b y$ are nonnegative (e.g., if A, x, and y are nonnegative), we have

$$\left[\text{sum}_{x,y}\left(A^{p+a+b}\right)\right]^{2k+1} \leq \left[\frac{\text{sum}_x(A^{2a}) + \text{sum}_y(A^{2b})}{2}\right]^{2k} \text{sum}_{x,y}\left(A^{p(2k+1)+a+b}\right) \qquad \text{and}$$

$$\left[\text{sum}_{x,y}\left(A^{p+a+b}\right)\right]^{2k} \leq \left[\frac{\text{sum}_x(A^{2a}) + \text{sum}_y(A^{2b})}{2}\right]^{2k-1} \frac{\text{sum}_x\left(A^{2kp+2a}\right) + \text{sum}_y\left(A^{2kp+2b}\right)}{2}.$$

The special case where $u = v$ and $a = b$ corresponds to Theorem 4.28.

Corollary 6.12

For any symmetric $n \times n$-matrix A, subsets $X, Y \subseteq [n]$, and all $a, b, k, p \in \mathbb{N}$ such that A^p, $A^a \psi(X)$, and $A^b \psi(Y)$ are nonnegative (e.g., if A is nonnegative), we have

$$\left[\text{sum}\left(A^{p+a+b}[X,Y]\right)\right]^{2k+1} \leq \left[\frac{\text{sum}(A^{2a}[X]) + \text{sum}(A^{2b}[Y])}{2}\right]^{2k} \text{sum}\left(A^{p(2k+1)+a+b}[X,Y]\right),$$

$$\left[\text{sum}\left(A^{p+a+b}[X,Y]\right)\right]^{2k} \leq \left[\frac{\text{sum}(A^{2a}[X]) + \text{sum}(A^{2b}[Y])}{2}\right]^{2k-1}.$$

$$\frac{\text{sum}\left(A^{2kp+2a}[X]\right) + \text{sum}\left(A^{2kp+2b}[Y]\right)}{2}.$$

The special case where $X = Y$ and $a = b$ corresponds to Corollary 4.10.

Corollary 6.13

For every undirected graph $G = (V, E)$, vertex subsets $X, Y \subseteq V$, and $a, b, k, p \in \mathbb{N}$, the following inequalies hold:

$$[w_{p+a+b}(X,Y)]^{2k+1} \leq \left[\frac{w_{2a}(X,X) + w_{2b}(Y,Y)}{2}\right]^{2k} \cdot w_{p(2k+1)+a+b}(X,Y) \qquad \text{and}$$

$$[w_{p+a+b}(X,Y)]^{2k} \leq \left[\frac{w_{2a}(X,X) + w_{2b}(Y,Y)}{2}\right]^{2k-1} \left[\frac{w_{2kp+2a}(X,X) + w_{2kp+2b}(Y,Y)}{2}\right].$$

The special case where $X = Y$ and $a = b$ corresponds to Corollary 4.11.

Corollary 6.14
For every undirected graph $G = (V, E)$, vertices $x, y \in V$, and $a, b, k, p \in \mathbb{N}$, the following inequalies hold:

$$[w_{p+a+b}(x,y)]^{2k+1} \leq \left[\frac{cl_{2a}(x) + cl_{2b}(y)}{2}\right]^{2k} \cdot w_{p(2k+1)+a+b}(x,y) \qquad and$$

$$[w_{p+a+b}(x,y)]^{2k} \leq \left[\frac{cl_{2a}(x) + cl_{2b}(y)}{2}\right]^{2k-1} \left[\frac{cl_{2kp+2a}(x) + cl_{2kp+2b}(y)}{2}\right] .$$

The special case where $x = y$ and $a = b$ corresponds to Corollary 4.13.

Again, we can "replace" the arithmetic mean $(\text{sum}_x(A^{2a}) + \text{sum}_y(A^{2b}))/2$ in Corollary 6.11 by the corresponding geometric mean if we use Theorem 6.9 or Corollary 6.8. To this end, we use the unit vectors

$$u := \frac{1}{\|A^a x\|} A^a x = \frac{1}{\sqrt{x^*(A^*)^a A^a x}} A^a x = \frac{1}{\sqrt{x^* A^{2a} x}} A^a x = \frac{1}{\sqrt{\text{sum}_x(A^{2a})}} A^a x$$

and

$$v := \frac{1}{\|A^b y\|} A^b y = \frac{1}{\sqrt{y^*(A^*)^b A^b y}} A^b y = \frac{1}{\sqrt{y^* A^{2b} y}} A^b y = \frac{1}{\sqrt{\text{sum}_y(A^{2b})}} A^b y .$$

Corollary 6.15
For any Hermitian $n \times n$-matrix A, vectors x, y, and $a, b, k, p \in \mathbb{N}$ such that A^p, $A^a x$, and $A^b y$ are nonnegative (e.g., if A, x, and y are nonnegative), we have

$$\left[\text{sum}_{x,y}(A^{p+a+b})\right]^{2k+1} \leq \left[\text{sum}_x(A^{2a}) \cdot \text{sum}_y(A^{2b})\right]^k \text{sum}_{x,y}(A^{p(2k+1)+a+b}) \qquad and$$

$$\left[\text{sum}_{x,y}(A^{p+a+b})\right]^{2k} \leq \frac{1}{2} \left[\text{sum}_x(A^{2a}) \cdot \text{sum}_y(A^{2b})\right]^k \left[\frac{\text{sum}_x(A^{2kp+2a})}{\text{sum}_x(A^{2a})} + \frac{\text{sum}_y(A^{2kp+2b})}{\text{sum}_y(A^{2b})}\right] .$$

The special case where $x = y$ and $a = b$ corresponds to Theorem 4.28.

Corollary 6.16
For any symmetric $n \times n$-matrix A, subsets $X, Y \subseteq [n]$, and $a, b, k, p \in \mathbb{N}$ such that A^p, $A^a \psi(X)$, and $A^b \psi(Y)$ are nonnegative (e.g., if A is nonnegative), we have

$$\left[\text{sum}(A^{p+a+b}[X,Y])\right]^{2k+1} \leq \left[\text{sum}(A^{2a}[X]) \cdot \text{sum}(A^{2b}[Y])\right]^k \text{sum}(A^{p(2k+1)+a+b}[X,Y]) ,$$

$$\left[\text{sum}(A^{p+a+b}[X,Y])\right]^{2k} \leq \frac{1}{2} \left[\text{sum}(A^{2a}[X]) \cdot \text{sum}(A^{2b}[Y])\right]^k \cdot$$
$$\left[\frac{\text{sum}(A^{2kp+2a}[X])}{\text{sum}(A^{2a}[X])} + \frac{\text{sum}(A^{2kp+2b}[Y])}{\text{sum}(A^{2b}[Y])}\right] .$$

The special case where $x = y$ and $a = b$ corresponds to Corollary 4.10.

Corollary 6.17
For every undirected graph $G = (V, E)$, all vertex subsets $X, Y \subseteq V$, and all $a, b, k, p \in \mathbb{N}$, the following inequalies hold:

$$[w_{p+a+b}(X,Y)]^{2k+1} \leq [w_{2a}(X,X) \cdot w_{2b}(Y,Y)]^k \cdot w_{p(2k+1)+a+b}(X,Y) \qquad and$$

$$[w_{p+a+b}(X,Y)]^{2k} \leq \frac{1}{2} [w_{2a}(X,X) \cdot w_{2b}(Y,Y)]^k \left[\frac{w_{2kp+2a}(X,X)}{w_{2a}(X,X)} + \frac{w_{2kp+2b}(Y,Y)}{w_{2b}(Y,Y)} \right] .$$

The special case where $X = Y$ and $a = b$ corresponds to Corollary 4.11.

If $p+a+b$ is an odd number, then the odd exponent case in the preceding corollaries leads to inequalities where a matrix power with *odd* exponent (or walk length) is also on the greater side, e.g., using $k = 1$, $p = 1$, $a = 0$, and $b = 2$, we have

$$w_3^3 \leq \left(\frac{w_0 + w_4}{2} \right)^2 \cdot w_5 \qquad \text{and} \qquad w_3^3 \leq w_0 \cdot w_4 \cdot w_5 .$$

Corollary 6.18
For every undirected graph $G = (V, E)$, vertices $x, y \in V$, and $a, b, k, p \in \mathbb{N}$, the following inequalies hold:

$$[w_{p+a+b}(x,y)]^{2k+1} \leq [cl_{2a}(x) \cdot cl_{2b}(y)]^k \cdot w_{p(2k+1)+a+b}(x,y) \qquad and$$

$$[w_{p+a+b}(x,y)]^{2k} \leq \frac{1}{2} [cl_{2a}(x) \cdot cl_{2b}(y)]^k \left[\frac{cl_{2kp+2a}(x)}{cl_{2a}(x)} + \frac{cl_{2kp+2b}(y)}{cl_{2b}(y)} \right] .$$

The special case where $x = y$ and $a = b$ corresponds to Corollary 4.13.

6.3.6   Generalizations for Complex Matrices

If we apply Lemma 6.2 together with Theorem 4.9, then we obtain the following result.

Theorem 6.10
For every complex $m \times n$-matrix A, vectors $u \in \mathbb{C}^m$ and $v \in \mathbb{C}^n$, and $a,b,c \in \mathbb{N}$, the following inequalities hold.
If c is odd, then we have

$$
\begin{aligned}
&\left[u^*(AA^*)^{a+\frac{c-1}{2}}Av + v^*(A^*A)^{a+\frac{c-1}{2}}A^*u\right] \cdot \\
&\left[u^*(AA^*)^{a+b+\frac{c-1}{2}}Av + v^*(A^*A)^{a+b+\frac{c-1}{2}}A^*u\right]
\end{aligned}
\leq
\begin{aligned}
&\left[u^*(AA^*)^a u + v^*(A^*A)^a v\right] \cdot \\
&\left[u^*(AA^*)^{a+b+c} u + v^*(A^*A)^{a+b+c} v\right] .
\end{aligned}
$$

If c is even, then we have

$$
\begin{aligned}
&\left[u^*(AA^*)^{a+\frac{c}{2}} u + v^*(A^*A)^{a+\frac{c}{2}} v\right] \cdot \\
&\left[u^*(AA^*)^{a+b+\frac{c}{2}} u + v^*(A^*A)^{a+b+\frac{c}{2}} v\right]
\end{aligned}
\leq
\begin{aligned}
&\left[u^*(AA^*)^a u + v^*(A^*A)^a v\right] \cdot \\
&\left[u^*(AA^*)^{a+b+c} u + v^*(A^*A)^{a+b+c} v\right] .
\end{aligned}
$$

Proof Consider the Hermitian matrix $B := \begin{pmatrix} 0 & A \\ A^* & 0 \end{pmatrix}$ and the vector s that is composed of the two vectors u and v. Applying Theorem 4.9 to B and s, we obtain

$$
s^* B^{2a+c} s s^* B^{2a+2b+c} s \leq s^* B^{2a} s s^* B^{2(a+b+c)} s .
$$

Now, the application of Lemma 6.2 yields the result.

If c is odd, the inequality corresponds to

$$
2\Re\left(u^*(AA^*)^{a+\frac{c-1}{2}}Av\right) \cdot 2\Re\left(u^*(AA^*)^{a+b+\frac{c-1}{2}}Av\right) \leq
$$
$$
\left[u^*(AA^*)^a u + v^*(A^*A)^a v\right] \cdot \left[u^*(AA^*)^{a+b+c} u + v^*(A^*A)^{a+b+c} v\right] .
$$

Corollary 6.19
For every complex $m \times n$-matrix A, all subsets $U \subseteq [m]$ and $V \subseteq [n]$, and all $a,b,c \in \mathbb{N}$, the following inequalities hold.
If c is odd, then we have

$$
\begin{aligned}
&\left[\mathrm{sum}\left(((AA^*)^{a+\frac{c-1}{2}}A)[U,V]\right) + \mathrm{sum}\left(((A^*A)^{a+\frac{c-1}{2}}A^*)[V,U]\right)\right] \cdot \\
&\left[\mathrm{sum}\left(((AA^*)^{a+b+\frac{c-1}{2}}A)[U,V]\right) + \mathrm{sum}\left(((A^*A)^{a+b+\frac{c-1}{2}}A^*)[V,U]\right)\right]
\end{aligned}
\leq
$$
$$
\begin{aligned}
&\left[\mathrm{sum}\left(((AA^*)^a)[U]\right) + \mathrm{sum}\left(((A^*A)^a)[V]\right)\right] \cdot \\
&\left[\mathrm{sum}\left(((AA^*)^{a+b+c})[U]\right) + \mathrm{sum}\left(((A^*A)^{a+b+c})[V]\right)\right] ,
\end{aligned}
$$

If c is even, then we have

$$
\begin{aligned}
&\left[\mathrm{sum}\left(((AA^*)^{a+\frac{c}{2}})[U]\right) + \mathrm{sum}\left(((A^*A)^{a+\frac{c}{2}})[V]\right)\right] \cdot \\
&\left[\mathrm{sum}\left(((AA^*)^{a+b+\frac{c}{2}})[U]\right) + \mathrm{sum}\left(((A^*A)^{a+b+\frac{c}{2}})[V]\right)\right]
\end{aligned}
\leq
$$
$$
\begin{aligned}
&\left[\mathrm{sum}\left(((AA^*)^a)[U]\right) + \mathrm{sum}\left(((A^*A)^a)[V]\right)\right] \cdot \\
&\left[\mathrm{sum}\left(((AA^*)^{a+b+c})[U]\right) + \mathrm{sum}\left(((A^*A)^{a+b+c})[V]\right)\right] .
\end{aligned}
$$

If c is odd, the inequality corresponds to

$$2\Re\left(\text{sum}\left(((AA^*)^{a+\frac{c-1}{2}}A)[U,V]\right)\right) \cdot 2\Re\left(\text{sum}\left(((AA^*)^{a+b+\frac{c-1}{2}}A)[U,V]\right)\right) \leq$$
$$\left[\text{sum}\left((AA^*)^a[U]\right)+\text{sum}\left((A^*A)^a[V]\right)\right]\left[\text{sum}\left((AA^*)^{a+b+c}[U]\right)+\text{sum}\left((A^*A)^{a+b+c}[V]\right)\right].$$

6.3.6.1 Implications for Hermitian Matrices

Corollary 6.20
For every Hermitian $n \times n$-matrix A, complex vectors $u, v \in \mathbb{C}^n$, and all $a, b, c \in \mathbb{N}$, the following inequalities hold.
If c is odd, then we have

$$\begin{aligned}\left[u^*A^{2a+c}v + v^*A^{2a+c}u\right]\cdot & \qquad \left[u^*A^{2a}u + v^*A^{2a}v\right]\cdot \\ \left[u^*A^{2a+2b+c}v + v^*A^{2a+2b+c}u\right] & \leq \left[u^*A^{2(a+b+c)}u + v^*A^{2(a+b+c)}v\right].\end{aligned}$$

If c is even, then we have

$$\begin{aligned}\left[u^*A^{2a+c}u + v^*A^{2a+c}v\right]\cdot & \qquad \left[u^*A^{2a}u + v^*A^{2a}v\right]\cdot \\ \left[u^*A^{2a+2b+c}u + v^*A^{2a+2b+c}v\right] & \leq \left[u^*A^{2(a+b+c)}u + v^*A^{2(a+b+c)}v\right].\end{aligned}$$

If c is odd, the inequality corresponds to

$$2\Re\left(u^*A^{2a+c}v\right)\cdot 2\Re\left(u^*A^{2a+2b+c}v\right) \leq \left[u^*A^{2a}u+v^*A^{2a}v\right]\cdot\left[u^*A^{2(a+b+c)}u+v^*A^{2(a+b+c)}v\right].$$

Corollary 6.21
For every Hermitian $n \times n$-matrix A, subsets $S, T \subseteq [n]$, and all $a, b, c \in \mathbb{N}$, the following inequalities hold.
If c is odd, then we have

$$\begin{aligned}\left[\text{sum}(A^{2a+c}[S,T]) + \text{sum}(A^{2a+c}[T,S])\right]\cdot & \qquad \left[\text{sum}(A^{2a}[S]) + \text{sum}(A^{2a}[T])\right]\cdot \\ \left[\text{sum}(A^{2a+2b+c}[S,T]) + \text{sum}(A^{2a+2b+c}[T,S])\right] & \leq \left[\text{sum}(A^{2(a+b+c)}[S]) + \text{sum}(A^{2(a+b+c)}[T])\right].\end{aligned}$$

If c is even, then we have

$$\begin{aligned}\left[\text{sum}(A^{2a+c}[S]) + \text{sum}(A^{2a+c}[T])\right]\cdot & \qquad \left[\text{sum}(A^{2a}[S]) + \text{sum}(A^{2a}[T])\right]\cdot \\ \left[\text{sum}(A^{2a+2b+c}[S]) + \text{sum}(A^{2a+2b+c}[T])\right] & \leq \left[\text{sum}(A^{2(a+b+c)}[S]) + \text{sum}(A^{2(a+b+c)}[T])\right].\end{aligned}$$

The special case where $S = T$ corresponds to Corollary 4.3.

Corollary 6.22

For every undirected graph $G = (V, E)$, vertex subsets $S, T \subseteq V$, and all $a, b, c \in \mathbb{N}$, the following inequalities hold.

If c is odd, then we have

$$w_{2a+c}(S,T) \cdot w_{2a+2b+c}(S,T) \leq \frac{w_{2a}(S,S) + w_{2a}(T,T)}{2} \cdot \frac{w_{2(a+b+c)}(S,S) + w_{2(a+b+c)}(T,T)}{2} .$$

If c is even, then we have

$$\begin{aligned} (w_{2a+c}(S,S) + w_{2a+c}(T,T)) \cdot \\ (w_{2a+2b+c}(S,S) + w_{2a+2b+c}(T,T)) \end{aligned} \leq \begin{aligned} (w_{2a}(S,S) + w_{2a}(T,T)) \cdot \\ (w_{2(a+b+c)}(S,S) + w_{2(a+b+c)}(T,T)) \end{aligned} .$$

Corollary 6.23

For every undirected graph $G = (V, E)$, all vertices $s, t \in V$, and all $a, b, c \in \mathbb{N}$, the following inequalities hold.

If c is odd, then we have

$$w_{2a+c}(s,t) \cdot w_{2a+2b+c}(s,t) \leq \frac{cl_{2a}(s) + cl_{2a}(t)}{2} \cdot \frac{cl_{2(a+b+c)}(s) + cl_{2(a+b+c)}(t)}{2} .$$

If c is even, then we have

$$\begin{aligned} (cl_{2a+c}(s) + cl_{2a+c}(t)) \cdot \\ (cl_{2a+2b+c}(s) + cl_{2a+2b+c}(t)) \end{aligned} \leq \begin{aligned} (cl_{2a}(s) + cl_{2a}(t)) \cdot \\ (cl_{2(a+b+c)}(s) + cl_{2(a+b+c)}(t)) \end{aligned} .$$

Chapter 7

Powers of Row and Column Sums

> If I feel unhappy, I do mathematics to become happy.
> If I am happy, I do mathematics to keep happy.
>
> *Alfréd Rényi*

A particularly important matrix class is the class of *nonnegative* matrices. For an overview, we refer to the books of Seneta [Sen73], Graham [Gra87], Minc [Min88], Berman and Plemmons [BP94] and Bapat and Raghavan [BR97].

7.1 Degree Powers and Walks in Directed Graphs

7.1.1 Related Work

Khintchine [Khi32][1] published the following inequality for $\{0,1\}$-matrices which was important to solve several problems in additive number theory (see Khintchine [Khi33] and Buchstab [Buc33]).

---

[1]The spelling exists in different ways: Khinchin, Khintchine, or Хинчин (Russian).

Theorem 7.1 (Khintchine)

Suppose that A is an $m \times n$ $\{0,1\}$-matrix with row sums r_i, $i \in [m]$, and column sums c_j, $j \in [n]$. Then we have

$$\sum_{i=1}^{m} r_i^2 + \sum_{j=1}^{n} c_j^2 \leq \mathrm{sum}(A) \left(\max\{m, n\} + \frac{\mathrm{sum}(A)}{\max\{m, n\}} \right) .$$

Luxemburg [Lux72] provided another proof and discussed extensions to bounded plane measurable sets.

Theorem 7.1 implies the following for directed graphs.

Corollary 7.1

For every directed graph with n vertices and m edges, we have

$$\sum_{x \in V} d_{in}(x)^2 + d_{out}(x)^2 \leq m(n + m/n) = w_1(w_0 + w_1/w_0) = m(n + \bar{d}) .$$

For *undirected* and for *directed Eulerian* graphs, we have $d_{in}(x) = d_{out}(x) = d(x)$ for all $x \in V$. Therefore $\sum_{x \in V} d_{in}(x)^2 + d_{out}(x)^2 = 2 \sum_{x \in V} d_{in}(x) \cdot d_{out}(x) = 2w_2$ implies $2w_2 \leq w_1(w_0 + w_1/w_0)$. Note that we have $w_1^2/w_0 \leq w_2 \leq w_1 w_0$ in this case.

In the *directed* case, we use $2w_2 \leq \sum_{x \in V} d_{in}(x)^2 + d_{out}(x)^2$ as follows (see also Corollary 7.6) to show a general statement.

Theorem 7.2

For every directed graph, we have

$$2w_2 \leq w_1(w_0 + w_1/w_0) = m(n + \bar{d}) .$$

In undirected graphs, this corresponds to $2w_2 \leq w_1(w_0 + w_1/w_0) = 2m(n + \bar{d})$.

Proof By the global decomposition lemma (see Corollary 1.1), the Cauchy-Schwarz inequality (see Theorem 1.1), and the AM-GM inequality (see Theorem 1.8), we have

$$w_2 = \sum_{x \in V} d_{in}(x) \cdot d_{out}(x) \leq \sqrt{\left(\sum_{x \in V} d_{in}(x)^2 \right) \left(\sum_{x \in V} d_{out}(x)^2 \right)} \leq \frac{1}{2} \sum_{x \in V} d_{in}(x)^2 + d_{out}(x)^2 .$$

The application of Corollary 7.1 completes the proof.

Matúš [Mat90] published the following result (see also Matúš and Tuzar [MT92]).

Theorem 7.3 (Matúš)
Suppose that A is an $n \times n$ $\{0,1\}$-matrix with row sums r_i and column sums c_i, $i \in [n]$. Then we have

$$\sum_{i=1}^{n} r_i^2 + \sum_{j=1}^{n} c_j^2 \;\leq\; n \cdot \text{sum}(A) + \sum_{i=1}^{n} r_i \cdot c_i \;\;.$$

Let us point out that this corresponds to the following statement.

Corollary 7.2
For every directed graph with n vertices and m edges, we have

$$\sum_{x \in V} d_{in}(x)^2 + d_{out}(x)^2 \;\leq\; mn + \sum_{x \in V} d_{in}(x) \cdot d_{out}(x) \;=\; w_1 w_0 + w_2 \;\;.$$

Tuzar [Tuz93] contributed a related inequality.

Theorem 7.4 (Tuzar)
For every directed graph, we have

$$mn - w_2 \;\leq\; (n^3 - n)/3 \;\;.$$

This corresponds to $w_1 w_0 - w_2 \leq (w_0^3 - w_0)/3$.

A generalization of Theorem 7.1 for real matrices was shown by van Dam [vDam98].

Theorem 7.5 (van Dam)
Suppose that A is an $m \times n$ real matrix with row sums r_i, $i \in [m]$, and column sums c_j, $j \in [n]$. Then we have

$$m \sum_{i=1}^{m} r_i^2 + n \sum_{j=1}^{n} c_j^2 \;\leq\; (\text{sum}(A))^2 + mn \sum_{i=1}^{m} \sum_{j=1}^{n} a_{i,j}^2 \;\;.$$

This is not only an improvement of Theorem 7.1, but also a generalization of the special case $\left(\sum_{i=1}^{n} x_i\right)^2 \leq n \sum_{i=1}^{n} x_i^2$ of Cauchy's inequality (see Theorem 1.1), which corresponds to the case of an $m \times 2$-matrix, where one of the two columns consists of the values x_i and the other column contains the values $-x_i$. Theorem 7.5 has been further extended by Feng and Tonge [FT10] to the following form using weighted row, column, and entry sums.

Theorem 7.6 (Feng and Tonge)
Suppose that A is an $m \times n$ real matrix and $x \in \mathbb{R}_{\geq 0}^m$ and $y \in \mathbb{R}_{\geq 0}^n$ are nonnegative vectors that satisfy $\sum_{i=1}^{m} x_i = \sum_{j=1}^{n} y_j = 1$. Then we have

$$\sum_{i=1}^{m} \left(\sum_{j=1}^{n} a_{i,j} y_j\right)^2 x_i + \sum_{j=1}^{n} \left(\sum_{i=1}^{m} a_{i,j} x_i\right)^2 y_j \;\leq\; (\text{sum}_{x,y}(A))^2 + \sum_{i=1}^{m} \sum_{j=1}^{n} a_{i,j}^2 x_i y_j \;\;.$$

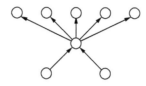

Figure 7.1: Example graph for $w_k \not\leq \sum_{x \in V} d_{\mathsf{in}}(x)^k$. **Setting** $k = 2$ **yields** $10 = w_2 \not\leq$ $\sum_{x \in V} d_{\mathsf{in}}(x)^2 = 9.$

7.1.2 Extended Results

In the following, we discuss possible generalizations of Theorem 4.32 to directed graphs and corresponding generalizations of Theorem 4.33 to nonsymmetric non-negative matrices.

The conceivable inequality $w_k \overset{?}{\leq} \sum_{x \in V} d_{\mathsf{in}}(x)^k$ is invalid. For instance, it is violated by the graph shown in Figure 7.1. Because of the reversely directed counterpart of this graph, the same applies to the inequality $w_k \overset{?}{\leq} \sum_{x \in V} d_{\mathsf{out}}(x)^k$. Also, trying to generalize the inequality by using direct products of $d_{\mathsf{in}}(x)$ and $d_{\mathsf{out}}(x)$ is not successful, since, e.g., $w_k \overset{?}{\leq} \sum_{x \in V} \sqrt{d_{\mathsf{in}}(x) \cdot d_{\mathsf{out}}(x)}^k$ is violated for $k = 1$ by the graph consisting of only one directed edge.

While the power sum for $d_{\mathsf{in}}(x)$ or $d_{\mathsf{out}}(x)$ alone is not suitable for bounding w_k, we will show that a combination (namely, the geometric mean) of both sums is sufficient. To this end, we first show that for the consideration of power sums with exponent q over the set of walks of length p the total cannot decrease if we shorten the walk length while at the same time the exponent is increased by the same amount.

Theorem 7.7

For all directed graphs $G = (V, E)$ and for all nonnegative integers $p, q \in \mathbb{N}$, the following inequality holds

$$\sum_{x \in V} d_{in}(x)^q s_p(x) \sum_{y \in V} d_{out}(y)^q e_p(y) \leq \sum_{x \in V} d_{in}(x)^{q+1} s_{p-1}(x) \sum_{y \in V} d_{out}(y)^{q+1} e_{p-1}(y) .$$

We will prove this theorem in the next subsection in a more general form for non-negative matrices (see Theorem 7.8).

Theorem 7.7 implies a chain of inequalities where the smallest and the largest elements are

$$\sum_{x \in V} s_{p+q}(x) \sum_{y \in V} e_{p+q}(y) = w_{p+q}^2 \quad \text{and} \quad \sum_{x \in V} d_{in}(x)^{p+q} \sum_{y \in V} d_{out}(y)^{p+q} .$$

Hence it directly implies the following corollary (see also the more general form in Corollary 7.5).

Corollary 7.3
For every directed graph $G = (V, E)$ and for all $k \in \mathbb{N}$, we have

$$w_k \leq \sqrt{\left(\sum_{v \in V} d_{in}(v)^k\right)\left(\sum_{v \in V} d_{out}(v)^k\right)} \; .$$

That means, although $w_k \not\leq \sum_{x \in V} d_{in}(x)^k$ and $w_k \not\leq \sum_{x \in V} d_{out}(x)^k$, we know for the geometric mean of the two power sums that

$$w_k \leq \mathfrak{G}\left(\sum_{v \in V} d_{in}(v)^k, \sum_{v \in V} d_{out}(v)^k\right) = \sqrt{\left(\sum_{x \in V} d_{in}(x)^k\right)\left(\sum_{x \in V} d_{out}(x)^k\right)} \; .$$

It is now natural to ask whether this kind of geometric mean bound could be improved to a harmonic mean bound, but this is not valid in general:

$$w_k \not\leq \mathfrak{H}\left(\sum_{v \in V} d_{in}(v)^k, \sum_{v \in V} d_{out}(v)^k\right) = \frac{2\left(\sum_{v \in V} d_{in}(v)^k\right)\left(\sum_{v \in V} d_{out}(v)^k\right)}{\sum_{v \in V} d_{in}(v)^k + \sum_{v \in V} d_{out}(v)^k} \; .$$

A counterexample for such a harmonic mean bound for directed graphs could be constructed in the same way like the graph in Figure 7.1. Construct a graph with $a + b + 1$ vertices and $a + b$ edges in the following way. There are b vertices with an edge to a central vertex, and there are a vertices with an edge from the central vertex. Then we have $w_2 = ab$, $\sum_{v \in V} d_{in}(v)^2 = b^2 + a$, and $\sum_{v \in V} d_{out}(v)^2 = a^2 + b$. Setting $a = 14$ and $b = 3$ yields $14 \cdot 3 = 42 \not\leq \frac{2(3^2 + 14)(14^2 + 3)}{14^2 + 14 + 3^2 + 3} = \frac{9154}{222} \approx 41.2$.

Corollary 7.3 implies the following statement by applying the inequality of arithmetic and geometric means.

Corollary 7.4
For every directed graph $G = (V, E)$ and for all $k \in \mathbb{N}$, we have

$$w_k \leq \frac{1}{2}\left(\sum_{v \in V} d_{in}(v)^k + d_{out}(v)^k\right).$$

Therefore, at least one of the two power sums must be greater than or equal to w_k:

$$w_k \leq \max\left\{\sum_{x \in V} d_{in}(x)^k, \sum_{x \in V} d_{out}(x)^k\right\}$$

Note that Corollaries 7.3 and 7.4 contain Theorem 4.32 by Fiol and Garriga as a special case (since $d_{in}(x) = d_{out}(x)$ holds for all x). Both corollaries could also be derived from corresponding inequalities for nonnegative matrices that were shown by Virtanen [Vir90] and Merikoski and Virtanen [MV95] (see Corollaries 7.6 and 7.5).

7.2 Row and Column Sums in Nonnegative Matrices

Now, we will generalize Theorem 4.33 and Theorem 7.7 to the case of arbitrary nonnegative matrices.

Theorem 7.8

For every nonnegative $n \times n$-matrix A with row sums r_i and column sums c_i, $i \in [n]$, and for all $p, q \in \mathbb{N}$, the following inequality holds:

$$\sum_{i \in [n]} c_i^q r_i^{[p]} \sum_{j \in [n]} r_j^q c_j^{[p]} \leq \sum_{i \in [n]} c_i^{q+1} r_i^{[p-1]} \sum_{j \in [n]} r_j^{q+1} c_j^{[p-1]} \ .$$

Proof

$$\sum_{i=1}^{n} c_i^q r_i^{[p]} \sum_{j=1}^{n} r_j^q c_j^{[p]} = \left(\sum_{i=1}^{n} c_i^q \sum_{j=1}^{n} a_{ij}^{[p]} \right) \left(\sum_{j=1}^{n} r_j^q \sum_{i=1}^{n} a_{ij}^{[p]} \right)$$

$$= \left(\sum_{i=1}^{n} c_i^q \sum_{j=1}^{n} \sum_{k=1}^{n} a_{ik}^{[p-1]} a_{kj} \right) \left(\sum_{j=1}^{n} r_j^q \sum_{i=1}^{n} \sum_{\ell=1}^{n} a_{i\ell} a_{\ell j}^{[p-1]} \right)$$

$$= \sum_{i=1}^{n} c_i^q \sum_{k=1}^{n} a_{ik}^{[p-1]} r_k \sum_{j=1}^{n} r_j^q \sum_{\ell=1}^{n} a_{\ell j}^{[p-1]} c_\ell = \sum_{i=1}^{n} \sum_{k=1}^{n} \sum_{j=1}^{n} \sum_{\ell=1}^{n} a_{ik}^{[p-1]} c_i^q r_k a_{\ell j}^{[p-1]} r_j^q c_\ell$$

$$= \sum_{(i,k) \in [n]^2} \left(\sum_{\substack{(\ell,j) \in [n]^2 \\ (\ell,j)=(i,k)}} a_{ik}^{[p-1]} a_{\ell j}^{[p-1]} c_i^q r_j^q r_k c_\ell + \sum_{\substack{(\ell,j) \in [n]^2 \\ (\ell,j) \neq (i,k)}} a_{ik}^{[p-1]} a_{\ell j}^{[p-1]} c_i^q r_j^q r_k c_\ell \right)$$

$$= \left(\sum_{(i,k) \in [n]^2} a_{ik}^{[p-1]^2} c_i^{q+1} r_k^{q+1} \right) + \sum_{(i,k) \in [n]^2} \sum_{\substack{(\ell,j) \in [n]^2 \\ (\ell,j) \neq (i,k)}} a_{ik}^{[p-1]} a_{\ell j}^{[p-1]} c_i^q r_j^q r_k c_\ell$$

$$= \left(\sum_{(i,k) \in [n]^2} a_{ik}^{[p-1]^2} c_i^{q+1} r_k^{q+1} \right) + \sum_{(i,k)<(\ell,j) \in [n]^4} a_{ik}^{[p-1]} a_{\ell j}^{[p-1]} \left(c_i^q r_j^q r_k c_\ell + c_\ell^q r_k^q r_j c_i \right)$$

$$\leq \left(\sum_{(i,k) \in [n]^2} a_{ik}^{[p-1]^2} c_i^{q+1} r_k^{q+1} \right) + \sum_{(i,k)<(\ell,j) \in [n]^4} a_{ik}^{[p-1]} a_{\ell j}^{[p-1]} \left(c_i^{q+1} r_j^{q+1} + c_\ell^{q+1} r_k^{q+1} \right)$$

$$= \sum_{i=1}^{n} \sum_{k=1}^{n} \sum_{\ell=1}^{n} \sum_{j=1}^{n} a_{ik}^{[p-1]} a_{\ell j}^{[p-1]} c_i^{q+1} r_j^{q+1} = \sum_{i=1}^{n} c_i^{q+1} \sum_{k=1}^{n} a_{ik}^{[p-1]} \sum_{j=1}^{n} r_j^{q+1} \sum_{\ell=1}^{n} a_{\ell j}^{[p-1]}$$

$$= \sum_{i=1}^{n} c_i^{q+1} r_i^{[p-1]} \sum_{j=1}^{n} r_j^{q+1} c_j^{[p-1]}$$

This theorem is not only a more detailed version of the following corollary which was already shown by Merikoski and Virtanen [MV95]. It also provides a short elementary proof for it.

Corollary 7.5 (Merikoski and Virtanen)
For every nonnegative $n \times n$-matrix A with row sums r_i and column sums c_i, $i \in [n]$, and all $p \in \mathbb{N}$, we have

$$\text{sum}(A^p) \leq \sqrt{\left(\sum_{i=1}^{n} c_i^p\right)\left(\sum_{i=1}^{n} r_i^p\right)} \ .$$

Proof   We show the squared form of the inequality. First, we notice that

$$\sum_{i \in [n]} c_i^0 r_i^{[p]} \sum_{j \in [n]} r_j^0 c_j^{[p]} = \sum_{i \in [n]} r_i^{[p]} \sum_{j \in [n]} c_j^{[p]} = (\text{sum}(A^p))^2 \ .$$

Hence, we start with the term on the left hand side and we apply Theorem 7.8 repeatedly until we end up with

$$\sum_{i \in [n]} c_i^0 r_i^{[p]} \sum_{j \in [n]} r_j^0 c_j^{[p]} \leq \cdots \leq \sum_{i \in [n]} c_i^p r_i^{[0]} \sum_{j \in [n]} r_j^p c_j^{[0]} = \sum_{i \in [n]} c_i^p \sum_{j \in [n]} r_j^p$$

The last equality is due the fact that $a_{ij}^{[0]}$ is 1 for $i = j$ and 0 otherwise, since A^0 is the identity matrix.

Corollary 7.5 generalizes Theorem 4.33 since r_i equals c_i for symmetric matrices. It also generalizes Corollary 7.3.

Since the right hand side of Corollary 7.5 can be interpreted as a geometric mean, the inequality of arithmetic and geometric means directly implies the following corollary, which was already proven by Virtanen [Vir90] using majorization. An alternative proof for the same result has been published by Merikoski and Virtanen [MV91].

Corollary 7.6 (Virtanen)
For every nonnegative $n \times n$-matrix A with row sums r_i and column sums c_i, $i \in [n]$, and all $p \in \mathbb{N}$, we have

$$\text{sum}(A^p) \leq \frac{1}{2}\left(\sum_{i=1}^{n} c_i^p + r_i^p\right) \ .$$

This is a more general form of Theorems 4.33 and 7.4.

7.3   Other Inequalities for the Number of Walks

Theorem 7.8 and Corollaries 7.5 and 7.6 imply Theorem 7.7 and Corollaries 7.3 and 7.4. Additionally, a more general form of Corollary 7.3 for walks is implied by Corollary 7.5.

Corollary 7.7
For every directed graph $G = (V, E)$ and for all $p \in \mathbb{N}$, we have

$$w_{pk} \leq \sqrt{\left(\sum_{v \in V} e_k(v)^p\right)\left(\sum_{v \in V} s_k(v)^p\right)} \, .$$

In contrast to undirected graphs, the number of walks in directed graphs can decrease for increasing walk lengths. In particular, the number of walks of length k can become 0 if the graph is cycle-free. Thus, inequalities similar to $w_{2a+c} \cdot w_{2a+2b+c} \leq w_{2a} \cdot w_{2(a+b+c)}$ cannot be true in general for directed graphs.

Corollary 1.1 and and the Cauchy-Schwarz inequality (see Theorem 1.1) imply that

$$\text{sum}(A^{k+\ell}) = \sum_{i \in [n]} c_i^{[k]} \cdot r_i^{[\ell]} \leq \sqrt{\sum_{i \in [n]} \left(c_i^{[k]}\right)^2} \sqrt{\sum_{i \in [n]} \left(r_i^{[\ell]}\right)^2} \, .$$

Using the following natural converse of Cauchy's inequality (see Steele [Ste04, p. 75])

$$\sqrt{\sum_{i=1}^n a_i^2} \sqrt{\sum_{i=1}^n b_i^2} \leq \frac{(m+M)/2}{\sqrt{mM}} \sum_{i=1}^n a_i b_i \qquad \text{for } m, M \in \mathbb{R} \text{ with } 0 < m \leq \frac{a_i}{b_i} \leq M,$$

we obtain

$$\text{sum}\left(A^{k+\ell}\right) \leq \sqrt{\sum_{i \in [n]} \left(c_i^{[k]}\right)^2} \sqrt{\sum_{i \in [n]} \left(r_i^{[\ell]}\right)^2} \leq \frac{(m+M)/2}{\sqrt{mM}} \text{sum}\left(A^{k+\ell}\right) \, .$$

if $m, M \in \mathbb{R}$ satisfy $0 < m \leq c_i^{[k]}/r_i^{[\ell]} \leq M < \infty$ for all $i \in [n]$. This provides a two-sided bound on $\text{sum}\left(A^{k+\ell}\right)$ for all matrices with strictly positive column sums of A^k and row sums of A^ℓ.

Theorem 7.9
Suppose that A is a real $n \times n$-matrix with $c_i^{[k]} > 0$ and $r_i^{[\ell]} > 0$ for all $i \in [n]$. If $0 < m \leq \min_{i \in [n]} c_i^{[k]}/r_i^{[\ell]}$ and $M \geq \max_{v \in V} c_i^{[k]}/r_i^{[\ell]}$ denote lower and upper bounds for the ratios of the column sums of A^k and the row sums of A^ℓ, then we have

$$\frac{\sqrt{mM}}{(m+M)/2} \sqrt{\sum_{i \in [n]} \left(c_i^{[k]}\right)^2} \sqrt{\sum_{i \in [n]} \left(r_i^{[\ell]}\right)^2} \leq \text{sum}\left(A^{k+\ell}\right) \leq \sqrt{\sum_{i \in [n]} \left(c_i^{[k]}\right)^2} \sqrt{\sum_{i \in [n]} \left(r_i^{[\ell]}\right)^2} \, .$$

This inequality can also be written as

$$\frac{\sqrt{mM}}{(m+M)/2} \left\|c^{[k]}\right\| \cdot \left\|r^{[\ell]}\right\| \leq \text{sum}\left(A^{k+\ell}\right) \leq \left\|c^{[k]}\right\| \cdot \left\|r^{[\ell]}\right\|$$

or

$$\frac{\sqrt{mM}}{(m+M)/2} \leq \frac{\text{sum}\left(A^{k+\ell}\right)}{\left\|c^{[k]}\right\| \cdot \left\|r^{[\ell]}\right\|} \leq 1 \, .$$

A direct consequence is the following corollary for directed graphs with strictly positive in- and out-degrees.

Corollary 7.8
Suppose that $G = (V, E)$ is a directed graph with $d_{in}(v) > 0$ and $d_{out}(v) > 0$ for all $v \in V$. If $0 < m \leq \min_{v \in V} e_k(v)/s_\ell(v)$ and $M \geq \max_{v \in V} e_k(v)/s_\ell(v)$ denote lower and upper bounds of the ratios $e_k(v)/s_\ell(v)$, then we have

$$\frac{\sqrt{mM}}{(m+M)/2} \sqrt{\sum_{v \in V} e_k(v)^2} \sqrt{\sum_{v \in V} s_\ell(v)^2} \leq w_{k+\ell} \leq \sqrt{\sum_{v \in V} e_k(v)^2} \sqrt{\sum_{v \in V} s_\ell(v)^2} \, .$$

For *undirected* graphs, this implies $\frac{\sqrt{mM}}{(m+M)/2} \sqrt{w_{2k} \cdot w_{2\ell}} \leq w_{k+\ell} \leq \sqrt{w_{2k} \cdot w_{2\ell}}$. if $m, M \in \mathbb{R}$ with $0 < m \leq w_k(v)/w_\ell(v) \leq M$ for all $v \in V$.

7.4 Geometric Means

Theorem 4.42 implies the following result for matrices with nonnegative row or column sums.

Corollary 7.9
For every matrix A with nonnegative row sums $r_i \in \mathbb{R}_{\geq 0}$ or nonnegative column sums $c_i \in \mathbb{R}_{\geq 0}$, $\bar{r} := \frac{1}{n} \sum_{i \in [n]} r_i$, and $\bar{c} := \frac{1}{n} \sum_{i \in [n]} c_i$, we have

$$\left(\prod_{i=1}^{n} r_i \right)^{\bar{r}} \leq \prod_{i=1}^{n} r_i^{r_i} \qquad or \qquad \left(\prod_{i=1}^{n} c_i \right)^{\bar{c}} \leq \prod_{i=1}^{n} c_i^{c_i} \, , \quad respectively.$$

Note that we have $\bar{r} = \bar{c} = \frac{1}{n} \text{sum}(A)$. Furthermore, the inequalities are equivalent to

$$\left(\prod_{i=1}^{n} r_i \right)^{1/n} \leq \left(\prod_{i=1}^{n} r_i^{r_i} \right)^{1/\sum_{i \in [n]} r_i} \qquad \text{and} \qquad \left(\prod_{i=1}^{n} c_i \right)^{1/n} \leq \left(\prod_{i=1}^{n} c_i^{c_i} \right)^{1/\sum_{i \in [n]} c_i} \, ,$$

respectively. Hence, they compare the geometric mean of the row (or column) sums with a weighted geometric mean of the row (or column) sums. If the corollary is applied to the ℓ-th power of the adjacency matrix of a digraph, then we obtain the following result.

Corollary 7.10
For every directed graph $G = (V, E)$ and $\ell \in \mathbb{N}$, we have

$$\left(\prod_{v \in V} s_\ell(v) \right)^{w_\ell/n} \leq \prod_{v \in V} s_\ell(v)^{s_\ell(v)} \qquad and \qquad \left(\prod_{v \in V} e_\ell(v) \right)^{w_\ell/n} \leq \prod_{v \in V} e_\ell(v)^{e_\ell(v)} \, .$$

Recall that the average degree in directed graphs was defined as $\bar{d} := \frac{1}{n} \sum_{v \in V} d_{in}(v) = \frac{1}{n} \sum_{v \in V} d_{out}(v) = m/n$. Then we obtain the following special case.

Corollary 7.11
For every directed graph $G = (V, E)$, we have

$$\left(\prod_{v \in V} d_{out}(v)\right)^{\bar{d}} \le \prod_{v \in V} d_{out}(v)^{d_{out}(v)} \quad and \quad \left(\prod_{v \in V} d_{in}(v)\right)^{\bar{d}} \le \prod_{v \in V} d_{in}(v)^{d_{in}(v)} \ .$$

These inequalities are equivalent to the geometric mean inequalities

$$\sqrt[n]{\prod_{v \in V} d_{\text{out}}(v)} \le \sqrt[m]{\prod_{v \in V} d_{\text{out}}(v)^{d_{\text{out}}(v)}} = \sqrt[m]{\prod_{(v,w) \in E} d_{\text{out}}(v)}$$

and

$$\sqrt[n]{\prod_{v \in V} d_{\text{in}}(v)} \le \sqrt[m]{\prod_{v \in V} d_{\text{in}}(v)^{d_{\text{in}}(v)}} = \sqrt[m]{\prod_{(u,v) \in E} d_{\text{in}}(v)} \ .$$

For the product of both inequalities, we obtain

$$\sqrt[n]{\prod_{v \in V} d_{\text{in}}(v) \cdot d_{\text{out}}(v)} \le \sqrt[m]{\prod_{v \in V} d_{\text{in}}(v)^{d_{\text{in}}(v)} \cdot d_{\text{out}}(v)^{d_{\text{out}}(v)}} = \sqrt[m]{\prod_{(u,v) \in E} d_{\text{in}}(v) \cdot d_{\text{out}}(u)} \ .$$

On the other hand, we have the following *invalid* variant

$$\sqrt[n]{\prod_{v \in V} d_{\text{in}}(v) \cdot d_{\text{out}}(v)} \not\le \sqrt[m]{\prod_{v \in V} d_{\text{in}}(v)^{d_{\text{out}}(v)} \cdot d_{\text{out}}(v)^{d_{\text{in}}(v)}} = \sqrt[m]{\prod_{(u,v) \in E} d_{\text{in}}(u) \cdot d_{\text{out}}(v)} \ .$$

A simple class of counterexamples can be constructed from directed bipartite graphs where $n/2$ vertices have edges to all other $n/2$ vertices, and additionally, there is for every vertex just one edge in the other direction. That means, for half of the vertices, we have $d_{\text{in}} = 1$ and $d_{\text{out}} = n/2$, for the other half we have $d_{\text{in}} = n/2$ and $d_{\text{out}} = 1$. Thus, the product $d_{\text{in}}(v)d_{\text{out}}(v)$ equals $n/2$ for every vertex v. The left side of the inequality is now equal to $n/2$ whereas the right side equals $\sqrt[m]{[(n/2)^1 \cdot 1^{n/2}]^n} = (n/2)^{n/m}$. This is less than $n/2$ since $n < m$ for $n > 2$.

Regarding the terms above, we know that

$$\prod_{v \in V} d_{\text{in}}(v)^{d_{\text{out}}(v)} \le \prod_{v \in V} d_{\text{out}}(v)^{d_{\text{out}}(v)} \quad and \quad \prod_{v \in V} d_{\text{out}}(v)^{d_{\text{in}}(v)} \le \prod_{v \in V} d_{\text{in}}(v)^{d_{\text{in}}(v)}$$

since $\sum_{v \in V} d_{\text{in}}(v) = m = \sum_{v \in V} d_{\text{out}}(v)$, see, e.g., Sándor and Debnath [SD00]. If both inequalities are multiplied, then we obtain the following relation for the two terms above:

$$\prod_{v \in V} d_{\text{in}}(v)^{d_{\text{out}}(v)} \cdot d_{\text{out}}(v)^{d_{\text{in}}(v)} \le \prod_{v \in V} d_{\text{in}}(v)^{d_{\text{in}}(v)} \cdot d_{\text{out}}(v)^{d_{\text{out}}(v)} \ .$$

APPLICATIONS

IV

Chapter 8

Bounds for the Largest Eigenvalue

If people do not believe that mathematics is simple,
it is only because they do not realize how complicated life is.

John von Neumann

8.1 Matrix Foundations

Suppose A is an $n \times n$-matrix of complex entries $a_{i,j}$. Then it has n (not necessarily distinct) eigenvalues $\lambda_i \in \mathbb{C}$, $i \in [n]$. The maximum modulus (absolute value) of these eigenvalues

$$\rho(A) := \max_{i \in [n]} |\lambda_i|$$

is called the *spectral radius* of A. This chapter is dedicated to connections between the spectral radius and the row and column sums of a matrix or its powers. In particular, we investigate the relations between the number of walks in a graph and the largest eigenvalue of its adjacency matrix.

Surveys on these subjects were written by Brualdi [Bru11] and Chatelin [Cha12]. Further information can be found, for example, in the books of Varga [Var00], Bapat and Raghavan [BR97], Parlett [Par98], and Knauer [Kna11].

8.1.1 Complex Matrices

Barankin [Bar45] obtained the following spectral radius bound for complex matrices using the absolute row and column sums of the matrix entries.

Theorem 8.1 (Barankin)
Let A be a complex $n \times n$ matrix with spectral radius $\rho(A)$. Then

$$\rho(A)^2 \leq \max_{i \in [n]} \left(\sum_{k=1}^{n} |a_{i,k}| \right) \left(\sum_{k=1}^{n} |a_{k,i}| \right) .$$

Compatible Norms.

A matrix norm $\|.\|_M$ is said to be *compatible* with a vector norm $\|.\|_V$ if $\|Ax\|_V \leq \|A\|_M \cdot \|x\|_V$ for all matrices A and all vectors x. For any vector norm $\|x\|_V$, an *induced norm* (or *operator norm*) $\|A\|_M$ can be defined in the following way:

$$\|A\|_M := \max_{\|x\|_V = 1} \|Ax\|_V = \max_{\|x\|_V \neq 0} \frac{\|Ax\|_V}{\|x\|_V}$$

Every matrix norm $\|.\|_M$ induced by a vector norm $\|.\|_V$ is compatible with $\|.\|_V$ by definition. Note that all induced norms are submultiplicative matrix norms.

Any submultiplicative matrix norm allows the following well-known upper bound for the spectral radius of a matrix.

Theorem 8.2
For every complex matrix A and any submultiplicative matrix norm $\|.\|$, the spectral radius of A is bounded by

$$\rho(A) \leq \|A\| \qquad and \qquad \rho(A) \leq \sqrt[k]{\|A^k\|} \qquad for\ k \in \mathbb{N} \setminus \{0\}.$$

Proof We reproduce the proof from Horn and Johnson [HJ13]. Assume that λ denotes an eigenvalue of A, $Ax = \lambda x$ with $|x| \neq 0$ (i.e., x is a corresponding eigenvector), and define the $n \times n$-matrix X such that all columns of X are equal to the eigenvector x. Then we have $AX = \lambda X$ and hence

$$|\lambda| \|X\| = \|\lambda X\| = \|AX\| \leq \|A\| \|X\| .$$

This means that the spectral radius obeys $|\lambda| \leq \rho(A) \leq \|A\|$. The second inequality follows from $\rho(A)^k = \rho(A^k) \leq \|A^k\|$.

Norms induced by L^1 and L^∞.

Let $\|x\|_1$ denote the L^1-vector or Manhattan norm, i.e., the sum of the absolute values of the vector entries. The resulting induced matrix norm is

$$\|A\|_1 = \max_{x:\ \|x\|_1 = 1} \|Ax\|_1 = \max_{x:\ \|x\|_1 \neq 0} \frac{\|Ax\|_1}{\|x\|_1} = \max_{j \in [n]} \sum_{i \in [n]} |a_{i,j}|$$

which is the maximum absolute column sum. Similarly, the L^∞-vector norm has as induced matrix norm the maximum absolute row sum of the matrix. Hence we have the following result:

$$\rho(A) \le \sqrt[k]{\max_{j \in [n]} \sum_{i \in [n]} |a_{i,j}^{[k]}|} \qquad \text{and} \qquad \rho(A) \le \sqrt[k]{\max_{i \in [n]} \sum_{j \in [n]} |a_{i,j}^{[k]}|} \ .$$

For nonnegative matrices, this means

$$\rho(A) \le \sqrt[k]{\max_{j \in [n]} c_j^{[k]}} \qquad \text{and} \qquad \rho(A) \le \sqrt[k]{\max_{i \in [n]} r_i^{[k]}} \ .$$

Euclidean / Spectral Norm.

For the Euclidean vector norm $\|v\|_2$, let

$$\|A\|_2 = \max_{x: \ \|x\| \ne 0} \frac{\|Ax\|_2}{\|x\|_2} = \max_{x: \ \|x\|_2 = 1} \|Ax\|_2$$

denote the corresponding induced matrix norm, i.e., the *spectral norm*. Note that $\|A\|_2 = \max_{i \in [n]} \sigma_i = \sqrt{\lambda_1(A^*A)}$, where σ_i refers to the singular values of A. Hence there is a close connection to alternating matrix powers. Similar to the case above, we have

$$\lambda_1 \le \|A\|_2 = \max_{\|x\|_2 = 1} \sqrt{\sum_{i=1}^n \left| \sum_{j=1}^n a_{i,j} x_j \right|^2} \le \max_{\|x\|_2 = 1} \sqrt{\sum_{i=1}^n \left(\sum_{j=1}^n |a_{i,j} x_j| \right)^2}$$

$$\le \max_{\|x\|_2 = 1} \sqrt{\sum_{i=1}^n \left(\sum_{j=1}^n |a_{i,j}|^2 \right) \left(\sum_{j=1}^n |x_j|^2 \right)} = \sqrt{\sum_{i=1}^n \sum_{j=1}^n |a_{i,j}|^2} \ .$$

Here, the last term is equal to the Frobenius norm $\|A\|_F$:

$$\|A\|_F = \sqrt{\sum_{i=1}^n \sum_{j=1}^n |a_{i,j}|^2} = \sqrt{\sum_{i=1}^n \sigma_i^2} \ .$$

This norm is sometimes also denoted by $\|A\|_2$ (although it is different from the spectral norm) since it is equal to the Schatten 2-norm. Other names are Schur norm or Hilbert-Schmidt norm.

Corollary 8.1
For every complex matrix A, we have

$$\rho(A) \le \sqrt[k]{\|A^k\|_2} \le \sqrt[k]{\|A^k\|_F} \ .$$

It is also known that $\|A\|_2 \le \sqrt{\|A\|_1 \|A\|_\infty}$, i.e., $\sigma_{\max}^2 \le (\max_{i \in [n]} c_i)(\max_{j \in [n]} r_j)$. This is improved for $m \times n$-matrices by the inequality of Schur [Sch11]:

$$\sigma_{\max}^2 \le \max_{i \in [m], j \in [n]} r_i c_j$$

In turn, this was improved for nonzero matrices by Nikiforov [Nik07b] to

$$\sigma_{\max}^2 \leq \max_{i \in [m]} \sum_{j \in [n]} |a_{ij}| c_j \leq \max_{a_{ij} \neq 0} r_i c_j$$

Sometimes, this is stronger than Schur's inequality, for instance in the case of the adjacency matrix of the star $K_{1,n}$. Here, Schur's inequality gives $\sigma^2 \leq n^2$, while the last inequality gives the best possible bound $\sigma^2 \leq n$.

For all complex matrices, any matrix norm $\|.\|$ that is compatible with a vector norm, and all $p \in \mathbb{N} \setminus \{0\}$, we have $\rho(A) \leq \|A^p\|^{1/p} \leq \|A\|$ and $\rho(A) = \lim_{p \to \infty} \|A^p\|^{1/p}$, see Gautschi [Gau53a; Gau53b] and Yamamoto [Yam67]. The following more general result can be found in Horn and Johnson [HJ13, Cor. 5.6.14].

Theorem 8.3
For every complex matrix A and any submultiplicative matrix norm $\|.\|$, we have

$$\rho(A) = \lim_{p \to \infty} \|A^p\|^{1/p} \ .$$

Now, we propose the following generalization of an inequality that occurs already in a paper of Das and Bapat [DB08]. There, it is credited to Horn and Johnson [HJ13] and restricted to Hermitian matrices and real vectors.

Theorem 8.4
For any matrix A with largest singular value $\sigma_1(A)$ and $x, y \in \mathbb{C}^n$, we have

$$|x^* A y| \leq \sigma_1(A) \sqrt{x^* x} \sqrt{y^* y} \ .$$

Proof By the Cauchy-Schwarz inequality (see Theorem 1.2), we have

$$|x^* A y| = |\langle x, Ay \rangle| \leq \sqrt{|\langle x, x \rangle|} \sqrt{|\langle Ay, Ay \rangle|} = \sqrt{x^* x} \sqrt{y^* A^* A y} \ .$$

Note that $A^* A$ is a positive-semidefinite Hermitian matrix, i.e., all eigenvalues are nonnegative real numbers. By the Rayleigh-Ritz Theorem (see Theorem 8.5), we know that

$$\frac{y^* A^* A y}{y^* y} \leq \lambda_1(A^* A) \qquad \text{or} \qquad \sqrt{y^* A^* A y} \leq \sqrt{\lambda_1(A^* A)} \sqrt{y^* y} \ ,$$

where $\lambda_1(A^* A)$ denotes the largest eigenvalue of the matrix $A^* A$. Then the proof is complete since the square roots of the eigenvalues of $A^* A$ are the *singular values* of A.

8.1.2   Normal Matrices

Normal matrices obey the following well-known equality.

Lemma 8.1
For any normal matrix A, the largest singular value equals the spectral radius, i.e.,

$$\sigma_1(A) = \rho(A) \ .$$

Proof   Since A is a normal matrix, there is a spectral decomposition $A = U^* \Lambda U$ with diagonal matrix Λ containing the eigenvalues of A, see Theorem 3.2. Since $A^* = (U^* \Lambda U)^* = U^* \Lambda^* U$, we have $A^* A = U^* \Lambda U U^* \Lambda^* U = U^* (\Lambda \Lambda^*) U$, i.e., the eigenvalues of $A^* A$ are the values $\lambda_i \bar{\lambda}_i$ for $i = \{1, \ldots n\}$. Those values are just the squares of the absolute value (modulus) of each eigenvalue. Thus we have $\sigma_1 = \sqrt{\lambda_1(A^* A)} = \max_{i \in [n]} |\lambda_i| = \rho(A)$.

From Theorem 8.4 and Lemma 8.1, we obtain the following Corollary.

Corollary 8.2
For any normal matrix A with spectral radius $\rho(A)$ and $x, y \in \mathbb{C}^n$, we have

$$|x^* A y| \leq \rho(A) \sqrt{x^* x} \sqrt{y^* y} .$$

8.1.3   Real Symmetric and Hermitian Matrices

The following famous result provides estimates for the largest and smallest eigenvalues of Hermitian matrices, see [HJ13].

Theorem 8.5 (Rayleigh-Ritz Theorem)
Suppose that A is a Hermitian matrix with eigenvalues $\lambda_1 \geq \ldots \geq \lambda_n$, then

$$\lambda_n x^* x \leq x^* A x \leq \lambda_1 x^* x \qquad \text{for all } x \in \mathbb{C}^n,$$

$$\lambda_1 = \max_{x \neq 0} \frac{x^* A x}{x^* x} = \max_{\|x\|=1} x^* A x \qquad and \qquad \lambda_n = \min_{x \neq 0} \frac{x^* A x}{x^* x} = \min_{\|x\|=1} x^* A x .$$

For $x = \mathbf{1}_n$, this theorem implies

$$\lambda_n \leq \frac{\mathbf{1}_n^T A \mathbf{1}_n}{\mathbf{1}_n^T \mathbf{1}_n} = \frac{\mathrm{sum}(A)}{n} \leq \lambda_1 .$$

The same principle can be applied to powers of A. The eigenvalues of A^k are the powers λ_i^k of A's eigenvalues λ_i. Thus, for the largest eigenvalue of A^k, we also have to take into account the value of λ_n^k if k is an even number. For real symmetric matrices, Hyyrö, Merikoski and Virtanen [HMV86] showed the following bound.

Theorem 8.6 (Hyyrö et al.)
For every nonzero real symmetric matrix A and $k \in \mathbb{N} \setminus \{0\}$, $\ell \in \mathbb{N}$, we have

$$\max\{\lambda_1, |\lambda_n|\} \geq \sqrt[k]{\frac{\mathrm{sum}\left(A^{2\ell+k}\right)}{\mathrm{sum}\left(A^{2\ell}\right)}} \geq \min_{i \in [n]} |\lambda_i| \qquad \textit{if } k \textit{ is even}$$

and

$$\lambda_1 \geq \sqrt[k]{\frac{\mathrm{sum}\left(A^{2\ell+k}\right)}{\mathrm{sum}\left(A^{2\ell}\right)}} \geq \lambda_n \qquad \textit{if } k \textit{ is odd.}$$

Actually, this can be extended to weighted entry sums for powers of Hermitian matrices.

Theorem 8.7

For every Hermitian matrix A, scaling vector $s \in \mathbb{C}^n$, and $k \in \mathbb{N} \setminus \{0\}$, $\ell \in \mathbb{N}$ such that $\mathrm{sum}_s(A^{2\ell}) > 0$, we have

$$\max\{\lambda_1, |\lambda_n|\} \geq \sqrt[k]{\frac{\mathrm{sum}_s\left(A^{2\ell+k}\right)}{\mathrm{sum}_s\left(A^{2\ell}\right)}} \geq \min_{i \in [n]} |\lambda_i| \qquad \text{if } k \text{ is even}$$

and

$$\lambda_1 \geq \sqrt[k]{\frac{\mathrm{sum}_s\left(A^{2\ell+k}\right)}{\mathrm{sum}_s\left(A^{2\ell}\right)}} \geq \lambda_n \qquad \text{if } k \text{ is odd.}$$

Proof   Any power A^k, $k \in \mathbb{N}$, of a Hermitian matrix A is Hermitian. By applying the Rayleigh-Ritz Theorem to A^k, we know that the largest and the smallest eigenvalue of A^k are

$$\lambda_1\left(A^k\right) = \max_{0 \neq x \in \mathbb{C}^n} \frac{x^* A^k x}{x^* x} \qquad \text{and} \qquad \lambda_n\left(A^k\right) = \min_{0 \neq x \in \mathbb{C}^n} \frac{x^* A^k x}{x^* x} \ .$$

The eigenvalues of A^k are the powers λ_i^k of the eigenvalues λ_i of A. Recall that the eigenvalues λ_i of a Hermitian matrix are real numbers. Thus, the largest eigenvalue of A^k is $\max\{\lambda_1, |\lambda_n|\}$ if k is even, and it is λ_1 if k is odd. The smallest eigenvalue of A^k is $\min_{i \in [n]} |\lambda_i|$ if k is even, and it is λ_n if k is odd. Let us remark that $\mathrm{sum}_s\left(A^{2\ell+k}\right)$ could be negative if k is odd, but $\mathrm{sum}_s\left(A^{2\ell}\right)$ and $\mathrm{sum}_s\left(A^{2\ell+k}\right)$ for even k are nonnegative since the corresponding matrices are positive-semidefinite. Thus, the k-th root is defined. The proof is completed by considering the particular vector $x = A^\ell s$ in the formulas above.

Now we want to show that the lower bounds of the spectral radius in Theorem 8.7 are monotonically increasing for the *even* exponents. Before, we show the monotonicity result for *all* exponents in the restricted case of positive-semidefinite matrices.

Theorem 8.8

For every positive-semidefinite matrix P, scaling vector $s \in \mathbb{C}^n$, and $k, \ell, x, y \in \mathbb{N}$, $k \geq 1$, such that $\mathrm{sum}_s(P^\ell) \neq 0$, we have

$$\sqrt[k]{\frac{\mathrm{sum}_s\left(P^{k+\ell}\right)}{\mathrm{sum}_s\left(P^\ell\right)}} \leq \sqrt[k+x]{\frac{\mathrm{sum}_s\left(P^{k+x+\ell+y}\right)}{\mathrm{sum}_s\left(P^{\ell+y}\right)}} \ .$$

Proof   The inequality is equivalent to

$$\left[\text{sum}_s\left(P^{k+\ell}\right)\right]^{k+x} \cdot \left[\text{sum}_s\left(P^{\ell+y}\right)\right]^k \leq \left[\text{sum}_s\left(P^\ell\right)\right]^{k+x} \cdot \left[\text{sum}_s\left(P^{k+x+\ell+y}\right)\right]^k .$$

First, we show that the bound is monotonically increasing with respect to increasing k, i.e., we consider

$$\left[\text{sum}_s\left(P^{k+\ell}\right)\right]^{k+x} \cdot \left[\text{sum}_s\left(P^\ell\right)\right]^k \leq \left[\text{sum}_s\left(P^\ell\right)\right]^{k+x} \cdot \left[\text{sum}_s\left(P^{k+x+\ell}\right)\right]^k$$

$$\left[\text{sum}_s\left(P^{k+\ell}\right)\right]^{k+x} \leq \left[\text{sum}_s\left(P^\ell\right)\right]^x \cdot \left[\text{sum}_s\left(P^{k+x+\ell}\right)\right]^k .$$

This inequality follows from Theorem 4.31. Now it is left to show that the bound is monotonically increasing with respect to increasing ℓ, i.e.,

$$\left[\text{sum}_s\left(P^{k+\ell}\right)\right]^k \cdot \left[\text{sum}_s\left(P^{\ell+y}\right)\right]^k \leq \left[\text{sum}_s\left(P^\ell\right)\right]^k \cdot \left[\text{sum}_s\left(P^{k+\ell+y}\right)\right]^k$$

$$\text{sum}_s\left(P^{k+\ell}\right) \cdot \text{sum}_s\left(P^{\ell+y}\right) \leq \text{sum}_s\left(P^\ell\right) \cdot \text{sum}_s\left(P^{k+\ell+y}\right) .$$

This inequality follows from Theorem 4.11.

Note that a special case of Theorem 8.8, namely for positive-semidefinite symmetric nonnegative matrices, was discussed in Marcus and Newman [MN62, p. 629].

Now we are prepared to prove that the spectral radius bounds in Theorem 8.7 are monotonically increasing with respect to both parameters if we restrict the set to the case of even exponents. This can also be considered as a generalization of the measure $\mu_{k\ell}$ in the paper of Hyyrö, Merikoski and Virtanen [HMV86].

Theorem 8.9
For every Hermitian matrix A, scaling vector $s \in \mathbb{C}^n$, and $k, \ell \in \mathbb{N}$ such that $\text{sum}_s(A^{2\ell}) \neq 0$, suppose that

$$F_{k,\ell}(A, s) := \sqrt[k]{\frac{\text{sum}_s\left(A^{2\ell+k}\right)}{\text{sum}_s\left(A^{2\ell}\right)}} = \sqrt[k]{\frac{s^* A^{2\ell+k} s}{s^* A^{2\ell} s}} .$$

Then the following inequality holds for $k, \ell, x, y \in \mathbb{N}$ with $k \geq 1$:

$$F_{2k,\ell}(A, s) \leq F_{2(k+x),\ell+y}(A, s) .$$

Proof   The claim corresponds to

$$\sqrt[2k]{\frac{\text{sum}_s(A^{2\ell+2k})}{\text{sum}_s(A^{2\ell})}} \leq \sqrt[2(k+x)]{\frac{\text{sum}_s(A^{2(\ell+y)+2(k+x)})}{\text{sum}_s(A^{2(\ell+y)})}}$$

Since A is Hermitian, we have $A = A^*$ and $A^2 = A^* A$. Now the equivalent condition for this positive-semidefinite matrix $P := A^2 = A^* A$ is

$$\sqrt[2k]{\frac{\text{sum}_s\left(P^{k+\ell}\right)}{\text{sum}_s\left(P^\ell\right)}} \leq \sqrt[2(k+x)]{\frac{\text{sum}_s\left(P^{k+x+\ell+y}\right)}{\text{sum}_s\left(P^{\ell+y}\right)}} .$$

All weighted entry sums are nonnegative since P is positive-semidefinite. After squaring both sides, the inequality corresponds to Theorem 8.8.

Theorem 8.9 implies that the bounds in Theorem 8.7 improve with increasing k and ℓ.

8.1.4 Nonnegative Matrices

Perron [Per07] proved the following theorem on matrices with strictly positive entries.

Theorem 8.10 (Perron)
A positive matrix A always has a real and positive eigenvalue r which is a simple root of the characteristic equation and exceeds the moduli of all the other eigenvalues. To this "maximal" eigenvalue r there corresponds an eigenvector z of A with positive coordinates $z_i > 0$ ($i \in [n]$).

This theorem was extended by Frobenius [Fro12] to the case of nonnegative matrices.

Theorem 8.11 (Frobenius)
A nonnegative matrix A always has a nonnegative eigenvalue r such that the moduli of all its eigenvalues do not exceed r. To this "maximal" eigenvalue there corresponds an eigenvector with nonnegative coordinates.
If A is also irreducible then r is positive, r is a simple root of the characteristic equation, and to r corresponds an eigenvector with positive coordinates.

The combination of these two theorems is called the Perron-Frobenius Theorem. For a nonnegative matrix A, the spectral radius $\rho(A)$ itself is an eigenvalue of A. Often, this "maximal" or "largest" eigenvalue $\lambda_1 = \rho(A)$ is called the Perron root of A.

For row and column sums in nonnegative matrices, the following theorem holds (see Gantmacher [Gan60] or Minc [Min88]).

Theorem 8.12
If A is a nonnegative matrix with maximal eigenvalue λ_1, row sums r_i, and column sums c_i for $i \in [n]$, then

$$\min_{i \in [n]} r_i \leq \lambda_1 \leq \max_{i \in [n]} r_i \qquad and \qquad \min_{i \in [n]} c_i \leq \lambda_1 \leq \max_{i \in [n]} c_i \ .$$

If A is irreducible, then equality holds on either side if and only if all row sums (or column sums, resp.) of A are equal.

A part of this statement is often credited to one of the papers of Frobenius [Fro08; Fro09; Fro12], although it seems that actually none of these papers contains such a statement explicitly.

Generalized bounds using row sums for the Perron root of partitioned nonnegative matrices were obtained by Deutsch and Wielandt [DW83].

Corollary 8.3
If A is any matrix such that A^k is nonnegative, then the spectral radius $\rho(A)$ is bounded by

$$\sqrt[k]{\min_{i \in [n]} c_i^{[k]}} \le \rho(A) \le \sqrt[k]{\max_{i \in [n]} c_i^{[k]}} \qquad and \qquad \sqrt[k]{\min_{i \in [n]} r_i^{[k]}} \le \rho(A) \le \sqrt[k]{\max_{i \in [n]} r_i^{[k]}} \ .$$

For a nonnegative matrix A with largest eigenvalue λ_1, this means

$$\sqrt[k]{\min_{i \in [n]} c_i^{[k]}} \le \lambda_1 \le \sqrt[k]{\max_{i \in [n]} c_i^{[k]}} \qquad and \qquad \sqrt[k]{\min_{i \in [n]} r_i^{[k]}} \le \lambda_1 \le \sqrt[k]{\max_{i \in [n]} r_i^{[k]}} \ .$$

Proof We apply Theorem 8.12 to the matrix power A^k. We exploit the fact that the eigenvalues of the matrix power A^k are the k-th powers of A's eigenvalues, i.e., λ_i^k. In particular, this implies that $\rho(A^k) = \rho(A)^k$.

The subsequent statement follows immediately from Theorem 8.1 and the Perron-Frobenius Theorem.

Corollary 8.4
Let A be any nonnegative matrix with largest eigenvalue λ_1. Then

$$\lambda_1^2 = \rho(A)^2 \le \max_{i \in [n]} r_i c_i \ .$$

Now we apply this corollary to a matrix power A^k.

Corollary 8.5
If A is any matrix such that A^k is nonnegative, then the spectral radius $\rho(A)$ is bounded by

$$\rho(A) \le \sqrt[2k]{\max_{i \in [n]} r_i^{[k]} c_i^{[k]}} \ .$$

For a nonnegative matrix A with largest eigenvalue λ_1, this means $\lambda_1 \le \sqrt[2k]{\max_{i \in [n]} r_i^{[k]} c_i^{[k]}}$.

Proof If we apply Corollary 8.4 to A^k, we obtain $\rho(A^k)^2 \le \max_{i \in [n]} r_i(A^k) c_i(A^k)$. As before, $\rho(A^k) = \rho(A)^k$ completes the proof of the first part. The second part is due to the Perron-Frobenius Theorem.

Loewy and London [LL78] proved an inequality for the eigenvalues of nonnegative matrices that is very similar to Lemma 1.4.

Theorem 8.13 (Loewy and London)
For every nonnegative $n \times n$ matrix with eigenvalues λ_i, $i \in [n]$ and for all $k, p \in \mathbb{N} \setminus \{0\}$, we have

$$\left(\sum_{i=1}^{n} \lambda_i^k \right)^p \le n^{p-1} \sum_{i=1}^{n} \lambda_i^{kp} \ .$$

On the one hand, if we compare this inequality to the preconditions of Lemma 1.4, it is interesting to see that it holds true despite the eigenvalues can be negative. On the other hand, it is not so surprising anymore if we see the proof of a slight extension.

Theorem 8.14
For every real $n \times n$ matrix with eigenvalues λ_i, $i \in [n]$ and for all $k, p \in \mathbb{N} \setminus \{0\}$ such that A^k is nonnegative, we have

$$\left(\sum_{i=1}^{n} \lambda_i^k\right)^p \leq n^{p-1} \sum_{i=1}^{n} \lambda_i^{kp} \qquad or \qquad \operatorname{tr}\left(A^k\right)^p \leq n^{p-1} \operatorname{tr}\left(A^{kp}\right) \ .$$

Proof Recall that the k-th moment of the eigenvalues corresponds to the trace of A^k, i.e., $\operatorname{tr}(A^k) = \sum_{i=1}^{n} a_{i,i}^{[k]} = \sum_{i=1}^{n} \lambda_i^k$. Now we use the condition that A^k is nonnegative. We apply Lemma 1.4 to $a_{i,i}^{[k]} \geq 0$, $i \in [n]$, and obtain

$$\left(\sum_{i=1}^{n} a_{i,i}^{[k]}\right)^p \leq n^{p-1} \sum_{i=1}^{n} \left(a_{i,i}^{[k]}\right)^p \leq n^{p-1} \sum_{i=1}^{n} a_{i,i}^{[kp]} \ .$$

The second inequality follows from the fact that $A^{kp} = (A^k)^p$. Thus, each main diagonal entry $a_{i,i}^{[kp]}$ of A^{kp} is calculated as a sum of $(a_{i,i}^{[k]})^p$ plus other nonnegative terms. Hence $(a_{i,i}^{[k]})^p \leq a_{i,i}^{[kp]}$ for all $i \in [n]$.

Obviously, this applies to nonnegative matrices, that is, Theorem 8.13 is a special case of it.

The following result can be found in the book of Minc [Min88, p. 27].

Theorem 8.15 (Minc)
Let A be a nonnegative matrix with nonzero row sums r_i ($i \in [n]$) and maximal eigenvalue λ_1. Then

$$\min_{i \in [n]} \left(\frac{1}{r_i} \sum_{t=1}^{n} a_{i,t} r_t\right) \leq \lambda_1 \leq \max_{i \in [n]} \left(\frac{1}{r_i} \sum_{t=1}^{n} a_{i,t} r_t\right) \ .$$

An analogous result holds for the column sums c_i.

Let us remark here that this corresponds to

$$\min_i \frac{r_i(A^2)}{r_i(A)} \leq \lambda_1 \leq \max_i \frac{r_i(A^2)}{r_i(A)} \qquad and \qquad \min_i \frac{c_i(A^2)}{c_i(A)} \leq \lambda_1 \leq \max_i \frac{c_i(A^2)}{c_i(A)}$$

or, more compactly, $\min_i r_i^{[2]}/r_i \leq \lambda_1 \leq \max_i r_i^{[2]}/r_i$ and $\min_i c_i^{[2]}/c_i \leq \lambda_1 \leq \max_i c_i^{[2]}/c_i$ under the required conditions. In the case of irreducible matrices, equality occurs if and only if all ratios are equal, see also Zhang and Li [ZL02, Cor. 2.5].

The theorem also implies a more general result.

Corollary 8.6
Let A be a nonnegative matrix with $r_i > 0$ for all $i \in [n]$ or $c_i > 0$ for all $i \in [n]$. Then the maximal eigenvalue λ_1 of A is bounded by

$$\min_i \frac{r_i(A^{2k})}{r_i(A^k)} \le \lambda_1^k \le \max_i \frac{r_i(A^{2k})}{r_i(A^k)} \quad or \quad \min_i \frac{c_i(A^{2k})}{c_i(A^k)} \le \lambda_1^k \le \max_i \frac{c_i(A^{2k})}{c_i(A^k)} \ ,$$

respectively.

Proof If A has strictly positive row (or column) sums, then the same is true for A^k. Thus, we can simply apply Theorem 8.15 to A^k. The equality $\lambda_1(A^k) = \rho(A^k) = \rho(A)^k = \lambda_1(A)^k$ for nonnegative matrices completes the proof.

Liu [Liu96] generalized Theorem 8.12 and Theorem 8.15 in the following way.

Theorem 8.16 (Liu)
If A is a nonnegative matrix without zero rows and $s \in \mathbb{N} \setminus \{0\}$, then

$$\min_{i \in [n]} \left(\frac{r_i\left(A^{k+s}\right)}{r_i\left(A^k\right)} \right)^{1/s} \le \lambda_1 \le \max_{i \in [n]} \left(\frac{r_i\left(A^{k+s}\right)}{r_i\left(A^k\right)} \right)^{1/s} .$$

An analogous result holds for the column sums c_i.

The inequality

$$\min_{i \in [n]} \sqrt{\sum_{j=1}^n \frac{a_{i,j} \sum_{k=1}^n a_{j,k} r_k}{r_i}} \le \lambda_1 \le \max_{i \in [n]} \sqrt{\sum_{j=1}^n \frac{a_{i,j} \sum_{k=1}^n a_{j,k} r_k}{r_i}}$$

that was shown for irreducible matrices by Zhang and Li [ZL02] is a special case of Theorem 8.16 since it corresponds to $\min_{i \in [n]} \sqrt{r_i(A^3)/r_i(A)} \le \lambda_1 \le \max_{i \in [n]} \sqrt{r_i(A^3)/r_i(A)}$.

Theorem 8.16 was generalized by Butler and Siegel [BS13] and Butler [But13].

Further References.

There exists a huge number of papers on results concerning eigenvalues of nonnegative matrices. For example, a whole series of papers with bounds on the Perron root of nonnegative matrices was published by Kolotilina [Kol93; Kol04; Kol05; Kol06]. In particular, it should be noted here that [Kol06] uses walks (called paths there) in the edge-weighted graph of a nonnegative matrix. Whole chapters devoted to the localization of the eigenvalues can be found in the books of Marcus and Minc [MM64] and Minc [Min88].

8.1.5 Symmetric Nonnegative Matrices

The following result for *symmetric* nonnegative matrices follows from Theorem 8.7 and the Perron-Frobenius Theorem (see Theorem 8.11).

Corollary 8.7
For every symmetric nonnegative matrix A, scaling vector $s \in \mathbb{C}^n$, and $k \in \mathbb{N} \setminus \{0\}$, $\ell \in \mathbb{N}$ such that $\text{sum}_s(A^{2\ell}) > 0$, we have

$$\sqrt[k]{\frac{\text{sum}_s\left(A^{2\ell+k}\right)}{\text{sum}_s\left(A^{2\ell}\right)}} \leq \lambda_1 \ .$$

This is connected to Theorem 8.16 by the observation that the simplified form with scaling vector $s = \mathbf{1}_n$ is the same as the sum of the row sums, i.e., $\text{sum}(A^k) = \sum_{i=1}^n r_i(A^k)$. Recall that these lower bounds are monotonically increasing for the subset of even k, see Theorem 8.9.

8.2 Graphs and Adjacency Matrices

Now we are going to discuss some relevant results in the context of directed and undirected graphs. It should be mentioned here that the Perron root (largest eigenvalue, spectral radius) of an adjacency matrix is sometimes called the *index* of the graph.

8.2.1 Directed Graphs

By Theorem 8.12, the following corollary is obvious, see also Cvetković, Doob and Sachs [CDS79, p. 83].

Corollary 8.8
The index λ_1 of any directed graph $G = (V, E)$ obeys the inequalities

$$\min_{v \in V} d_{in}(v) \leq \lambda_1 \leq \max_{v \in V} d_{in}(v) \qquad \text{and} \qquad \min_{v \in V} d_{out}(v) \leq \lambda_1 \leq \max_{v \in V} d_{out}(v) \ .$$

A consequence of Corollary 8.3 is the following generalization of Corollary 8.8.

Corollary 8.9
The index λ_1 of any directed graph $G = (V, E)$ obeys the inequalities

$$\sqrt[k]{\min_{v \in V} e_k(v)} \leq \lambda_1 \leq \sqrt[k]{\max_{v \in V} e_k(v)} \qquad \text{and} \qquad \sqrt[k]{\min_{v \in V} s_k(v)} \leq \lambda_1 \leq \sqrt[k]{\max_{v \in V} s_k(v)} \ .$$

If we apply Corollary 8.5 to adjacency matrices, we obtain the following bound relating the index of the graph to the number of k-step walks starting at fixed vertices.

Corollary 8.10
In every directed graph $G = (V, E)$, the index λ_1 is bounded by

$$\lambda_1 \le \sqrt[k]{\max_{v \in V} \sqrt{e_k(v) s_k(v)}} = \sqrt[2k]{\max_{v \in V} e_k(v) s_k(v)} \ .$$

This means that in any directed graph, the index λ_1 is bounded from above by the $2k$-th root of the maximum number (over all $v \in V$) of $2k$-step walks with fixed central vertex v.

For the adjacency matrix A of any directed graph, all entries $a_{i,j}$ are either 0 or 1. The same applies to the terms $|a_{i,j}|^2$ that appear in the Frobenius norm of A. Therefore, Corollary 8.1 implies the following result.

Corollary 8.11
For all directed graphs with m edges, we have

$$\lambda_1 \le \sqrt{m} \ .$$

Other results related to the number of edges and the index of a directed graph were obtained, for example, by Schwarz [Sch64], Brualdi and Hoffman [BH85], and Friedland [Fri85].

Theorem 8.14 implies the following inequalities for the total number of closed walks.

Corollary 8.12
In all directed graphs, the closed walks fulfill the inequalities

$$cl_k^p \le n^{p-1} cl_{kp} \qquad or \qquad \left(\frac{cl_k}{n} \right)^p \le \frac{cl_{kp}}{n} \ .$$

This is very similar to the closed walks result in Corollary 4.13, but there only one vertex is considered.

A lower bound for the index of directed graphs using closed 2-step walks (i.e., anti-parallel edges) was shown by Gudiño and Rada [GR10]:

$$\lambda_1 \ge \frac{cl_2}{n} \ .$$

By applying Theorem 8.15 to the adjacency matrix A of a directed graph, we conclude the following result for walks starting from (or ending at) a certain vertex. Consider a directed graph where either all out-degrees are nonzero or all in-degrees are nonzero. Then, the index λ_1 obeys

$$\min_{v \in V} \frac{s_2(v)}{d_{\text{out}}(v)} \leq \lambda_1 \leq \max_{v \in V} \frac{s_2(v)}{d_{\text{out}}(v)} \qquad \text{or} \qquad \min_{v \in V} \frac{e_2(v)}{d_{\text{in}}(v)} \leq \lambda_1 \leq \max_{v \in V} \frac{e_2(v)}{d_{\text{in}}(v)} \ ,$$

respectively. Actually, we can state this in a more general way.

Corollary 8.13
Suppose that $G = (V, E)$ is a directed graph such that $d_{out}(v) > 0$ for all $v \in V$ or $d_{in}(v) > 0$ for all $v \in V$. Then we have

$$\min_{v \in V} \frac{s_{2k}(v)}{s_k(v)} \leq \lambda_1^k \leq \max_{v \in V} \frac{s_{2k}(v)}{s_k(v)} \qquad or \qquad \min_{v \in V} \frac{e_{2k}(v)}{e_k(v)} \leq \lambda_1^k \leq \max_{v \in V} \frac{e_{2k}(v)}{e_k(v)}$$

respectively.

Proof   Note that $d_{\text{out}}(v) > 0$ for all $v \in V$ implies that $s_k(v) > 0$ for all $v \in V$ and $k \in \mathbb{N}$. Similarly, $d_{\text{in}}(v) > 0$ for all $v \in V$ implies that $e_k(v) > 0$ for all $v \in V$ and $k \in \mathbb{N}$. Hence, we can simply apply Theorem 8.15 to the adjacency matrix power A^k. Again, using the equality $\lambda_1(A^k) = \rho(A^k) = \rho(A)^k = \lambda_1(A)^k$ for nonnegative matrices A completes the proof.

Xu, Fang and Shen [XFS12] obtained the following result.

Theorem 8.17 (Xu, Fang and Shen)
If G is a directed graph without loops and multi-edges such that $d_{out}(v) > 0$ for all $v \in V$, then

$$\min_{v \in V} \frac{s_{k+r}(v)}{s_k(v)} \leq \lambda_1^r \leq \max_{v \in V} \frac{s_{k+r}(v)}{s_k(v)} \ .$$

Actually, it is possible to obtain this generalization of Corollaries 8.8 and 8.13 directly from Theorem 8.16.

Corollary 8.14
Suppose that $G = (V, E)$ is a directed graph such that $d_{out}(v) > 0$ for all $v \in V$ or $d_{in}(v) > 0$ for all $v \in V$. Then the index λ_1 of G is bounded by

$$\min_{v \in V} \frac{s_{k+r}(v)}{s_k(v)} \leq \lambda_1^r \leq \max_{v \in V} \frac{s_{k+r}(v)}{s_k(v)} \qquad or \qquad \min_{v \in V} \frac{e_{k+r}(v)}{e_k(v)} \leq \lambda_1^r \leq \max_{v \in V} \frac{e_{k+r}(v)}{e_k(v)} \ ,$$

respectively.

Further results in this direction can be found in the article of Butler and Siegel [BS13] and in the thesis of Butler [But13].

Kwapisz [Kwa96] investigated the relations between the index, the in- and out-degrees, and the number of k-step walks of a directed graph. For the next theorem, let W_k denote the set of all k-step walks and for any k-step walk $w \in W_k$, let $v_0(w)$ and $v_k(w)$ denote the start and end vertex of w, respectively.

Theorem 8.18 (Kwapisz)
If $G = (V, E)$ is a directed graph with index λ_1, then the following estimates hold:

$$\lambda_1 \leq d_G^{(-1)} := \max_{(x,y) \in E} \left\{ \sqrt{d_{in}(y) \cdot d_{out}(x)} \right\}$$

$$\lambda_1 \leq d_G^{(k)} := \max_{w \in W_k} \left\{ \sqrt{d_{in}(v_0(w)) \cdot d_{out}(v_k(w))} \right\} , \quad k \in \mathbb{N}.$$

Moreover, Kwapisz showed that

$$w_k(x, y) \leq \left(d_G^{(-1)} \right)^k .$$

Also, for $e, f \in E$, the number of k-step walks starting with e and ending with f does not exceed $(d_G^{(0)})^{k-1}$. Further, for $k \in \mathbb{N}$, $\ell \geq k$, and any two k-step walks $\alpha, \beta \in W_k$, the number of ℓ-step walks starting with α and ending with β does not exceed $(d_G^{(k)})^{\ell-k}$. Note that $d_G^{(-1)}$ refers to the number of alternating 3-step walks through fixed edges and that d_G^k for $k \in \mathbb{N}$ considers the different numbers of $(k+2)$-walks through fixed k-step walks (making up the middle part, extended at the beginning and at the end by single edges). In particular, for $k = 0$, the term considers the numbers of 2-step walks through fixed vertices. Note also that the special case $\lambda_1 \leq d_G^{(0)} = \max_{x \in V} \{ \sqrt{d_{in}(x) \cdot d_{out}(x)} \}$ follows already from Theorem 8.1 and its Corollary 8.4. The special case of $d_G^{(-1)}$ and $d_G^{(1)}$ for *undirected* graphs is

$$\lambda_1 \leq \max_{\{x,y\} \in E} \left\{ \sqrt{d_x d_y} \right\} ,$$

which was later also obtained by Berman and Zhang [BZ01]. A similar bound can be deduced using the Randić index of undirected graphs:

$$M_{\mathrm{R}} = \sum_{\{x,y\} \in E} \sqrt{d_x d_y} .$$

More precisely, the arithmetic mean $\frac{1}{m} \sum_{\{x,y\} \in E} \sqrt{d_x d_y}$ is a lower bound of λ_1, see Van Mieghem [Van11, p. 47]. These two inequalities generalize the classical two-sided bound $\overline{d} \leq \lambda_1 \leq \Delta$.

Further results on spectra of digraphs can be found in the survey of Brualdi [Bru10].

8.2.2 Directed Graphs with Normal Adjacency Matrix

Theorem 8.19
For every directed graph on n vertices and m edges that has a normal adjacency matrix with largest eigenvalue λ_1, we have

$$\overline{d} = \frac{m}{n} \leq \lambda_1 .$$

Proof For any *normal* matrix A, the numerical radius satisfies $r(A) = \rho(A) = \|A\|_2$, see [HJ13, Problem 21, p. 331]. For general matrices, we have $r(A) \le \|A\|_2$. If A is a nonnegative normal matrix, then

$$\lambda_1 = \rho(A) = r(A) = \max_{x \ne 0} \left| \frac{x^* A x}{x^* x} \right| \ge \frac{\mathbf{1}_n^T A \mathbf{1}_n}{\mathbf{1}_n^T \mathbf{1}_n} = \frac{\text{sum}(A)}{n} .$$

For adjacency matrices, the last term corresponds to m/n.

Since the powers of nonnegative normal matrix are nonnegative and normal again, we can generalize the preceding result to the following form.

Theorem 8.20

For every directed graph having a normal adjacency matrix with largest eigenvalue λ_1, we have

$$\max_{S \subseteq V, |S| > 0} \sqrt[k]{\frac{w_k(S)}{|S|}} \le \lambda_1 .$$

Proof If A is a normal adjacency matrix, then

$$\lambda_1^k = \rho(A)^k = \rho(A^k) = r(A^k) = \max_{x \ne 0} \left| \frac{x^* A^k x}{x^* x} \right| \ge \frac{\psi(S)^T A^k \psi(S)}{\psi(S)^T \psi(S)} = \frac{w_k(S)}{|S|} .$$

As we have seen in the proof of Lemma 8.1, the eigenvalues of $B^* B$ are the values $\lambda_i \bar{\lambda}_i$ for $i = \{1, \dots n\}$. If B is a Hermitian matrix, then all the eigenvalues λ_i are real and the eigenvalues of $B^* B$ are just the squared eigenvalues of B. Then the largest eigenvalue of $B^* B$ is just $\max\{\lambda_1^2, \lambda_n^2\}$. If B is a nonnegative matrix, then we know (by the Perron-Frobenius Theorem) that the largest eigenvalue equals the spectral radius. If we consider a directed graph $G = (V, E)$ with normal adjacency matrix A and two vertex sets $X, Y \subseteq V$, we can set $B = A^r$, $x = (A^a)^T \psi(X)$, and $y = A^b \psi(Y)$. Then we can apply Corollary 8.2 for B, x, and y to obtain the following specialized form.

Corollary 8.15

For all directed graphs with normal adjacency matrix, we have

$$\frac{w_{a+r+b}(X, Y)}{\sqrt{\text{alt}_{2a}(X, X) \cdot \text{alt}_{2b}(Y, Y)}} \le \rho(A)^r = \lambda_1^r .$$

Let us remark that it does not make a difference for the number of alternating walks whether we start with a forward step or a backward step since $(A^* A)^k = (AA^*)^k$ for normal matrices A (i.e., $(A^T A)^k = (AA^T)^k$ if A is normal and real). If we use $X = Y = V$, we obtain the special case that uses the total number of alternating walks.

Corollary 8.16
For all directed graphs with normal adjacency matrix, we have

$$\frac{w_{a+r+b}}{\sqrt{\mathrm{alt}_{2a} \cdot \mathrm{alt}_{2b}}} \leq \rho(A)^r = \lambda_1^r .$$

8.2.3 Undirected Graphs

Collatz and Sinogowitz [CS57] noticed that the Rayleigh-Ritz Theorem can be applied to the adjacency matrices of undirected graphs. As a consequence, the average degree is a lower bound for the index of the graph:

$$\overline{d} = \frac{2m}{n} \leq \lambda_1 .$$

This is a special case of the observation $\lambda_n \leq \mathrm{sum}(A)/n \leq \lambda_1$ (see the comment after Theorem 8.5). Using the same principle, Hofmeister [Hof88] (and later also Favaron, Mahéo and Saclé [FMS93] and Zhou [Zho00]) showed that

$$\frac{1}{n} \sum_{v \in V} d_v^2 \leq \lambda_1^2 .$$

Further eigenvalue bounds in undirected graphs were investigated, for instance, by Wilf [Wil67] ($\lambda_1 \leq \sqrt{2m(1 - 1/n)}$), Brigham and Dutton [BD84], Brualdi and Hoffman [BH85], Friedland [Fri85; Fri88], Stanley [Sta87] ($\lambda_1 \leq (-1 + \sqrt{1 + 8m})/2$), and Hong [Hon93]. Several other publications considered the sum of squares of walk numbers to obtain the lower bounds

$$\sum_{v \in V} w_2(v)^2 / \sum_{v \in V} d_v^2 \;\leq\; \lambda_1^2 \qquad \text{Yu, Lu and Tian [YLT04],}$$

$$\sum_{v \in V} w_3(v)^2 / \sum_{v \in V} w_2(v)^2 \;\leq\; \lambda_1^2 \qquad \text{Hong and Zhang [HZ05],}$$

$$\sum_{v \in V} w_{k+1}(v)^2 / \sum_{v \in V} w_k(v)^2 \;\leq\; \lambda_1^2 \qquad \text{Hou, Tang and Woo [HTW07] .}$$

The further special case $\sum_{v \in V} w_4(v)^2 / \sum_{v \in V} w_3(v)^2 \leq \lambda_1^2$ was later published by Hu [Hu09]. Apparently, it has not been noticed at that time that all those bounds can be expressed more concisely using the total number of walks of the double length. While the bounds of Collatz and Sinogowitz and Hofmeister are equivalent to

$$\frac{w_1}{w_0} \leq \lambda_1 \qquad \text{and} \qquad \frac{w_2}{w_0} \leq \lambda_1^2 ,$$

the other bounds can be expressed as

$$\frac{w_4}{w_2} \leq \lambda_1^2 , \qquad \frac{w_6}{w_4} \leq \lambda_1^2 , \qquad \frac{w_8}{w_6} \leq \lambda_1^2 , \qquad \text{and} \qquad \frac{w_{2k+2}}{w_{2k}} \leq \lambda_1^2 .$$

These results were generalized by Nikiforov [Nik06b] as follows.[1]

---

[1] Note that Nikiforov defined the length of a walk as the number of vertices instead of the number of edges in his paper.

Theorem 8.21 (Nikiforov)

For every undirected graph G and for all $k \in \mathbb{N}$ and $r \in \mathbb{N} \setminus \{0\}$, the index λ_1 of G is bounded by

$$\frac{w_{2k+r}}{w_{2k}} \leq \lambda_1^r .$$

Note that this inequality also follows from the result of Hyyrö, Merikoski and Virtanen [HMV86], see Theorem 8.6. Nikiforov additionally characterized the case of equality. The inequality was later deduced again in the book of Van Mieghem [Van11, (3.38), p. 48]. In particular, Theorem 8.21 implies a bound using the average number of walks of length k and a bound regarding the growth factor for odd / even walk lengths:

$$\frac{w_r}{n} \leq \lambda_1^r \qquad \text{and} \qquad \frac{w_{2\ell+1}}{w_{2\ell}} \leq \lambda_1$$

which also contains the bound of Collatz and Sinogowitz as a special case.

From Corollary 8.16, we obtain the following more general form of Theorem 8.21.

Corollary 8.17

For all undirected graphs and $a, b, r \in \mathbb{N}$, we have

$$\frac{w_{a+r+b}}{\sqrt{w_{2a} w_{2b}}} \leq \lambda_1^r .$$

Setting $a = b = k$ leads to $\frac{w_{2k+r}}{w_{2k}} \leq \lambda_1^r$, i.e., Nikiforov's inequality (Theorem 8.21). Note that all these lower bounds for λ_1 can be used together with the inequality $\lambda_1 \leq (1 - 1/\omega)n$ of Wilf [Wil86] to obtain relations between the number of walks and the clique number ω of the graph. In particular, together with the inequality $\frac{2m}{n} \leq \lambda_1$, the concise Turán theorem $m \leq (1 - 1/\omega)n^2/2$ is obtained. Some of these connections are discussed in the survey of Nikiforov [Nik11] and in a recent paper of Ning [Nin14].

Note that the inequality of Wilf was extended by Nikiforov [Nik02; Nik09] to $\lambda_1^2 \leq 2(1 - 1/\omega)m$, which had been conjectured by Edwards and Elphick [EE83]. Nikiforov [Nik06b] also showed $\lambda_1^k \leq (1 - 1/\omega)w_k$.

Abdo, Dimitrov, Réti and Stevanović [ADRS14] showed experimental evidence that $\sqrt{M_2/m}$ is a good estimate for λ_1. Let us remark that this is not too surprising since it is equal to $\sqrt{w_3/w_1}$ and, in an asymptotic view, the development from A^k to A^{k+1} is dictated by $\rho(A) = \lambda_1$.

The following upper bound for the index of undirected graphs using walks starting at certain vertices was published by Nikiforov [Nik06b].

Corollary 8.18 (Nikiforov)

For all undirected graphs $G = (V, E)$, $k \in \mathbb{N}$, and $r \in \mathbb{N} \setminus \{0\}$, we have

$$\lambda_1^r \leq \max_{v \in V} \frac{w_{k+r}(v)}{w_k(v)} .$$

This was generalized to directed graphs by Xu, Fang and Shen [XFS12], see Theorem 8.17. Nikiforov's bound is a special case of the following two-sided bound that follows from Theorem 8.16 and Corollary 8.14.

Corollary 8.19
In all undirected graphs, we have

$$\min_{v \in V} \frac{w_{k+r}(v)}{w_k(v)} \leq \lambda_1^r \leq \max_{v \in V} \frac{w_{k+r}(v)}{w_k(v)} \ .$$

Further information on index bounds in undirected graphs can be found in the survey of Cvetković and Rowlinson [CR90] and in the book of Van Mieghem [Van11].

8.2.4 Other Bounds

As a special case for adjacency matrices of undirected graphs, we obtain the following graph index bounds from Corollary 8.7.

Corollary 8.20
For every nonempty undirected graph $G = (V, E)$, $k \in \mathbb{N} \setminus \{0\}$, and $\ell \in \mathbb{N}$, the index λ_1 satisfies the inequality

$$\lambda_1 \geq \max_{S \subseteq V, w_\ell(S) > 0} \sqrt[k]{\frac{w_{2\ell+k}(S,S)}{w_{2\ell}(S,S)}} \ .$$

The single vertex case $S = \{v\}$ corresponds to the following closed walks form.

Corollary 8.21
For every nonempty undirected graph $G = (V, E)$, $k \in \mathbb{N} \setminus \{0\}$, and $\ell \in \mathbb{N}$, we have

$$\lambda_1 \geq \max_{v \in V, w_\ell(v) > 0} \sqrt[k]{\frac{cl_{2\ell+k}(v)}{cl_{2\ell}(v)}} \qquad \text{and (for } \ell = 0) \qquad \lambda_1 \geq \max_{v \in V} \sqrt[k]{cl_k(v)} \ .$$

That means, Corollaries 8.20 and 8.21 are generalizations of the lower bound $\lambda_1 \geq \sqrt{\Delta}$ by Nosal [Nos70] that uses the maximum degree Δ (since $w_2(v, v) = d(v)$). Recall also that, due to Theorem 8.9, the lower bounds for the graph index in Corollaries 8.20 and 8.21 are monotonically increasing for the subset of even walk lengths (assuming $w_{2\ell}(S, S) > 0$):

$$\sqrt[k]{\frac{w_{2(k+\ell)}(S,S)}{w_{2\ell}(S,S)}} \leq \sqrt[k+x]{\frac{w_{2(k+x+\ell+y)}(S,S)}{w_{2(\ell+y)}(S,S)}} \ .$$

Corollary 4.11 directly implies the additional monotonicity result

$$\sqrt[p]{\frac{w_{2\ell+p}(S,S)}{w_{2\ell}(S,S)}} \leq \sqrt[pk]{\frac{w_{2\ell+pk}(S,S)}{w_{2\ell}(S,S)}} \ .$$

Again, the most interesting special cases are $S = \{v\}$ for closed walks and $S = V$ for all walks, in view of Theorem 8.21. In contrast to the statement that follows from Theorem 8.9, these inequalities provide subset monotonicity statements also for certain *odd* walk lengths.

8.3 Eigenvalue Moments and Energy

An application for the lower bounds of the graph index are upper bounds for the sum of the absolute values of all eigenvalues of the adjacency matrix. This is an important measure in theoretical chemistry. In particular, it appears in the Hückel molecular orbital (HMO) method, or short Hückel method. This semi-empiric method in quantum chemistry tries to approximate molecular orbitals of π-electrons in conjugated hydrocarbon systems by a simple linear combination of atomic orbitals (LCAO). It predicts the number of energy levels (and also which levels are degenerate in the sense that they correspond to two or more different states of the quantum system) and it approximates the molecular orbital energies as a function of the Coulomb and interaction energy parameters α and β. Despite its simplifying assumptions it yields surprisingly good results. Let us start with a brief review of the bounds obtained by McClelland [McC71]. Assuming that the orbital energy for the i-th molecular orbital is $\alpha + \lambda_i \beta$, then the Hückel molecular orbital energy of n π-electrons in a conjugated system of n atoms is $E_\pi = n\alpha + E\beta$ with $E = \sum_{i=1}^{n} m_i \lambda_i$, where $0 \leq m_i \leq 2$ is the occupation number of the i-th molecular orbital. The values λ_i are the eigenvalues of the adjacency matrix of the conjugated system. Now the interesting case appears to be $E = 2 \sum_{i:\ \lambda_i > 0} \lambda_i$, where bonding orbitals are doubly occupied and antibonding orbitals are empty. Since the trace of the matrix is $\mathrm{tr}(A) = 0 = \sum_{i=1}^{n} \lambda_i$, we have $E = \sum_{i=1}^{n} |\lambda_i|$. This motivated the definition of *graph energy* (see Gutman [Gut78]) for an undirected graph G as

$$E(G) := \sum_{i=1}^{n} |\lambda_i| \ .$$

The fact that the singular values of normal matrices are equal to the absolute values of the eigenvalues suggests to extend the concept of graph energy to the energy of a matrix by considering the sum of its singular values. Indeed, this has been proposed by Nikiforov [Nik07c; Nik07a]. For a recent survey on graph energy and matrix norms (in particular Ky Fan and Schatten norms, i.e., generalizations of the trace norm), see the recent paper by Nikiforov [Nik15].

8.3.1 Energy Bounds

First bounds for the energy of undirected graphs were published by McClelland [McC71]:

$$\sqrt{2m + n(n-1)|\det A|^{2/n}} \leq E(G) \leq \sqrt{2mn}.$$

The upper bound is an immediate consequence of the Cauchy-Schwarz inequality (applied to $\mathbf{1}_n$ and the vector containing the eigenvalues):

$$E(G) = \sum_{i=1}^{n} 1 \cdot |\lambda_i| \leq \sqrt{\sum_{i=1}^{n} 1^2} \sqrt{\sum_{i=1}^{n} \lambda_i^2} = \sqrt{n \cdot cl_2} = \sqrt{n \cdot 2m} \ .$$

For the lower bound, McClelland used

$$\left(\sum_{i=1}^{n} |\lambda_i| \right)^2 = \sum_{i=1}^{n} |\lambda_i|^2 + \sum_{(i,j) \in [n]^2, i \neq j} |\lambda_i||\lambda_j|$$

and then he applied the inequality of arithmetic and geometric means to the last term:

$$\frac{1}{n(n-1)} \sum_{(i,j) \in [n]^2, i \neq j} |\lambda_i||\lambda_j| \geq \left(\prod_{(i,j) \in [n]^2, i \neq j} |\lambda_i||\lambda_j| \right)^{1/n(n-1)} = \left(\prod_{i \in [n]} |\lambda_i| \right)^{2/n}$$

$$= |\det A|^{2/n} \ .$$

Other early investigations are due to Gutman and Trinajstić [GT72; GT73]. Later, numerous other bounds were published, for a survey see Gutman [Gut01]. For more recent surveys on graph energy, see Gutman [Gut05], Gutman, Li and Zhang [GLZ09], Li, Shi and Gutman [LSG12], and Triantafillou [Tri14].

Koolen and Moulton [KM01] reported the following upper bound on the graph energy. Since $\lambda_1 \geq 0$, we have $E(G) = \lambda_1 + \sum_{i=2}^{n} |\lambda_i|$. By application of the Cauchy-Schwarz inequality, we obtain $\sum_{i=2}^{n} |\lambda_i| \leq \sqrt{(n-1)\sum_{i=2}^{n} \lambda_i^2}$. Additionally, we can apply the equality $\sum_{i=1}^{n} \lambda_i^2 = 2m$ for simple undirected graphs. Hence, the graph energy can be written as

$$E(G) = \lambda_1 + \sum_{i=2}^{n} |\lambda_i| \leq \lambda_1 + \sqrt{(n-1)\sum_{i=2}^{n} \lambda_i^2} = \lambda_1 + \sqrt{(n-1)(2m - \lambda_1^2)} \ .$$

More generally, Lemma 1.4 implies $\sum_{i=2}^{n} |\lambda_i| \leq \sqrt[2k]{(n-1)^{2k-1} \sum_{i=2}^{n} \lambda_i^{2k}}$ for $k \in \mathbb{N} \setminus \{0\}$. Thus, we have

$$E(G) = \lambda_1 + \sum_{i=2}^{n} |\lambda_i| \leq \lambda_1 + \sqrt[2k]{(n-1)^{2k-1} \sum_{i=2}^{n} \lambda_i^{2k}} = \lambda_1 + \sqrt[2k]{(n-1)^{2k-1}(cl_{2k} - \lambda_1^{2k})} \ .$$

In bipartite graphs, we have $\lambda_1 = -\lambda_n$ and the inequality $\sum_{i=2}^{n-1} |\lambda_i| \leq \sqrt{(n-2)\sum_{i=2}^{n-1} \lambda_i^2}$ can be used to obtain (see Koolen, Moulton and Gutman [KMG00] and Koolen and Moulton [KM03])

$$E(G) = 2\lambda_1 + \sum_{i=2}^{n-1} |\lambda_i| \leq 2\lambda_1 + \sqrt{(n-2)\sum_{i=2}^{n-1} \lambda_i^2} \leq 2\lambda_1 + \sqrt{(n-2)(2m - 2\lambda_1^2)} \ .$$

Therefore, all general graph energy bounds that are described in the subsequent subsection have a similar counterpart for bipartite graphs. Based on the preceding observations, Koolen and Moulton [KM01] showed the following result (see also [KMG00]).

Theorem 8.22 (Koolen and Moulton)
If $2m \geq n$ and G is a graph on n vertices with m edges, then

$$E(G) \leq \frac{2m}{n} + \sqrt{(n-1)\left[2m - \left(\frac{2m}{n}\right)^2\right]} \; .$$

Hofmeister's bound $\frac{1}{n}\sum_{v \in V} d_v^2 \leq \lambda_1^2$ [Hof88; FMS93; Zho00] was used by Zhou [Zho04] to show the following.

Theorem 8.23 (Zhou)
If $G = (V, E)$ is a graph with n vertices and m edges, then

$$E(G) \leq \sqrt{\frac{\sum_{v \in V} d(v)^2}{n}} + \sqrt{(n-1)\left(2m - \frac{\sum_{v \in V} d(v)^2}{n}\right)} \; .$$

Using the inequality $\sum_{v \in V} w_2(v)^2 / \sum_{v \in V} d_v^2 \leq \lambda_1^2$ [YLT04], Yu, Lu and Tian [YLT05] obtained the result

$$E(G) \leq \sqrt{\frac{\sum_{v \in V} w_2(v)^2}{\sum_{v \in V} d(v)^2}} + \sqrt{(n-1)\left(2m - \frac{\sum_{v \in V} w_2(v)^2}{\sum_{v \in V} d(v)^2}\right)} \; .$$

Using the inequality $\sum_{v \in V} w_3(v)^2 / \sum_{v \in V} w_2(v)^2 \leq \lambda_1^2$ [HZ05], Liu, Lu and Tian [LLT07] obtained the result

$$E(G) \leq \sqrt{\frac{\sum_{v \in V} w_3(v)^2}{\sum_{v \in V} w_2(v)^2}} + \sqrt{(n-1)\left(2m - \frac{\sum_{v \in V} w_3(v)^2}{\sum_{v \in V} w_2(v)^2}\right)} \; .$$

A generalized upper bound was shown by Hou, Tang and Woo [HTW07] using their lower bound for the graph index $\sum_{v \in V} w_{k+1}(v)^2 / \sum_{v \in V} w_k(v)^2 \leq \lambda_1^2$.

Theorem 8.24 (Hou, Tang and Woo)
The energy of a connected undirected graph G with $n \geq 2$ vertices and m edges is bounded by

$$E(G) \leq \sqrt{\frac{\sum_{v \in V} w_{k+1}(v)^2}{\sum_{v \in V} w_k(v)^2}} + \sqrt{(n-1)\left(2m - \frac{\sum_{v \in V} w_{k+1}(v)^2}{\sum_{v \in V} w_k(v)^2}\right)} \; .$$

We prefer to express this in a different way that seems to be more natural.

Corollary 8.22
The energy of a connected undirected graph G with $n \geq 2$ vertices and m edges is bounded by

$$E(G) \leq \sqrt{\frac{w_{2k+2}}{w_{2k}}} + \sqrt{(n-1)\left(2m - \frac{w_{2k+2}}{w_{2k}}\right)} \ .$$

For a review of those bounds, see also Li, Shi and Gutman [LSG12, pp. 61–65].

We now deduce a generalized upper bound for the graph energy, using our lower bounds for the spectral radius.

Theorem 8.25
Suppose that $G = (V, E)$ is an undirected graph. For every vertex subset $S \subseteq V$ such that the induced subgraph $G[S]$ has average degree $\overline{d}(G[S]) \geq \overline{d}(G) = 2m/n$, we have

$$E(G) \leq \sqrt[2k]{\frac{w_{2k+2\ell}(S,S)}{w_{2\ell}(S,S)}} + \sqrt{(n-1)\left(2m - \sqrt[k]{\frac{w_{2k+2\ell}(S,S)}{w_{2\ell}(S,S)}}\right)} \ .$$

Proof The function $f(x) = x + \sqrt{(n-1)(2m - x^2)}$ has derivative $f'(x) = 1 - \frac{\sqrt{n-1}x}{\sqrt{2m-x^2}}$ and is therefore monotonically decreasing in the interval $\sqrt{2m/n} \leq x \leq \sqrt{2m}$. Hence, we have

$$\sqrt{2m} \geq \lambda_1 \geq F_{k,\ell}(S) \geq F_{1,0}(S) = \sqrt{\frac{w_2(S,S)}{w_0(S,S)}} \geq \sqrt{\frac{w_1(S,S)}{w_0(S,S)}} = \sqrt{\frac{|E(G[S])|}{|S|}}.$$

Thus, we have $f(\lambda_1) \leq f(F_{k,\ell}(S))$ for every nonempty set S inducing a subgraph $G[S]$ with average degree $\overline{d}(G[S]) \geq \overline{d} = 2m/n$. For each such set S, this implies

$$E(G) \leq f(\lambda_1) \leq f(F_{k,\ell}(S)) \leq \sqrt[2k]{\frac{w_{2k+2\ell}(S,S)}{w_{2\ell}(S,S)}} + \sqrt{(n-1)\left(2m - \sqrt[k]{\frac{w_{2k+2\ell}(S,S)}{w_{2\ell}(S,S)}}\right)}.$$

Since $\frac{w_2(S,S)}{w_0(S,S)} \geq \frac{w_1(S,S)^2}{w_0(S,S)^2}$ (Corollary 4.4), the same applies if $\frac{w_1(S,S)}{w_0(S,S)} = \overline{d}(G[S]) \geq \sqrt{\frac{2m}{n}}$.

8.3.2 Closed Walks Bounds

By Lemma 1.4, we have (for $p \geq 1$)

$$\left(\sum_{i=1}^{n} |\lambda_i|\right)^p \leq n^{p-1} \sum_{i=1}^{n} |\lambda_i|^p \ .$$

Thus, we conclude

$$E(G) = \sum_{i=1}^{n} |\lambda_i| \leq \sqrt[p]{n^{p-1} \sum_{i=1}^{n} |\lambda_i|^p} = n\sqrt[p]{\frac{1}{n}\sum_{i=1}^{n}|\lambda_i|^p} \qquad \text{for } p \geq 1.$$

Since the trace of A^ℓ ($\ell \in \mathbb{N}$) is the number of closed walks of length ℓ, that is, we have $\sum_{i=1}^{n} \lambda_i^\ell = \sum_{v \in V} w_\ell(v, v)$ and since $\sum_{i=1}^{n} |\lambda_i|^{2k} = \sum_{i=1}^{n} \lambda_i^{2k}$ for $k \in \mathbb{N}$, we conclude

$$E(G) = \sum_{i=1}^{n} |\lambda_i| \leq n \sqrt[2k]{\frac{1}{n} \sum_{i=1}^{n} \lambda_i^{2k}} = n \sqrt[2k]{\frac{1}{n} \sum_{v \in V} w_{2k}(v, v)}$$

Theorem 8.26
For $k \in \mathbb{N}$, the graph energy and the number of closed walks obey the following bound:

$$E(G) \leq n \sqrt[2k]{\frac{cl_{2k}}{n}} \ .$$

The special case $k = 1$ with $cl_2 = 2m$ corresponds to the upper bound $E(G) \leq \sqrt{2mn}$ of McClelland [McC71]. A related upper bound for the graph energy is discussed in the paper of Li [Li09].

Chapter 9

Iterated Kernels

This system is quite unlike the case of mathematics, in which everything can be defined, and then we do not know what we are talking about. In fact, the glory of mathematics is that we do not have to say what we are talking about. The glory is that the laws, the arguments, and the logic are independent of what 'it' is.

Richard P. Feynman

9.1 Related Work

Atkinson, Watterson and Moran [AWM60] remarked that Theorem 6.4 is equivalent to

$$\left(\sum_{i=1}^{m} \sum_{j=1}^{n} a_{ij} \right)^3 \leq mn \sum_{i=1}^{m} \sum_{j=1}^{n} a_{ij} \sum_{r=1}^{m} a_{rj} \sum_{s=1}^{n} a_{is}$$

and that this implies the following analogous integral form: If the nonnegative function $K(x, y) \geq 0$ is integrable on the rectangle $0 \leq x \leq a, 0 \leq y \leq b$, then

$$\left(\int_0^a \int_0^b K(x, y) \, dx \, dy \right)^3 \leq ab \int_0^a \int_0^b \int_0^a \int_0^b K(x, t) \, K(x, y) \, K(s, y) \, dx \, dy \, ds \, dt \ .$$

If $a = b$ and K is symmetric, that is $K(x, y) = K(y, x)$, this yields

$$\left(\int_0^a \int_0^a K(x, y) \, dx \, dy \right)^3 \leq a^2 \int_0^a \int_0^a \int_0^a \int_0^a K(t, x) \, K(x, y) \, K(y, s) \, dx \, dy \, ds \, dt$$

If an *iterated kernel* for a square-integrable function K on $0 \leq x, y \leq a$ is defined as

$$K_1(x, y) = K(x, y) \qquad \text{and} \qquad K_{i+1}(x, y) = \int_0^a K_i(x, s) \, K(s, y) \, ds \ ,$$

then this can be written more concisely as

$$\left(\int_0^a \int_0^a K(x,y)\, dx\, dy \right)^3 \leq a^2 \int_0^a \int_0^a K_3(x,y)\, dx\, dy \;.$$

On the one hand, this kind of iteration only works for square integrable versions, that is, if $a = b$. On the other hand, there is also another definition of iterated kernels which does not require $a = b$, but which exchanges the order of parameters in every second step. In a natural way, this corresponds to alternating matrix powers and we call it an *alternating iterated kernel*. For example, if $K(x,y)$ is considered as an analog of a matrix A, then $H(x,y) = \int_a^b K(x,s)\,K(y,s)\,ds$ is an analog of AA^T, see also Horn and Johnson [HJ91, p. 138]. If $K(x,y)$ has complex function values, we assume that, additionally to switching the parameters, we also use the complex conjugate $\overline{K(y,x)}$ in every second step. This corresponds to going from $(AA^T)^k[A]$ to $(AA^*)^k[A]$. While the alternating iterated kernels correspond to alternating powers of rectangular matrices, the "straight" iterated kernels correspond to normal powers of square matrices. To distinguish these two types, we use $K_{(i)}$ for the alternating version instead of K_i. Similar to the case of Hermitian matrices and alternating matrix powers, there is no difference between straight and alternating iterated kernels if K is Hermitian, that is, if $K(x,y) = \overline{K(y,x)}$.

For the case $\ell = 2^r 3^s$ $(r, s \in \mathbb{N})$ of nonnegative symmetric kernels $K(x,y) = K(y,x) \geq 0$, Atkinson et al. proved that

$$\left(\int_0^a \int_0^a K(x,y)\, dx\, dy \right)^\ell \leq a^{\ell-1} \int_0^a \int_0^a K_\ell(x,y)\, dx\, dy \;,$$

but they made the conjecture that the inequality is true for all integers ℓ. Beesack [Bee69] noticed that this more general form is already implied by the result of Mulholland and Smith [MS59] (see Theorem 4.20). More generally, Beesack notes that if v is any nonnegative function, then

$$\left(\int_0^a \int_0^a v(x)\, v(y)\, K(x,y)\, dx\, dy \right)^\ell \leq \left(\int_0^a v^2(x) \right)^{\ell-1} \int_0^a \int_0^a v(x)\, v(y)\, K_\ell(x,y)\, dx\, dy$$

This follows by replacing the integrals by approximating Riemann sums in the form of Theorem 4.20 (and considering the limits for $k \to \infty$):

$$\left(v^T A v \right)^\ell \leq \left(v^T v \right)^{\ell-1} \left(v^T A^\ell v \right)$$

$$\left(\sum_{i=1}^k \sum_{j=1}^k v_i a_{ij} v_j \right)^\ell \leq \left(\sum_{i=1}^k v_i^2 \right)^{\ell-1} \sum_{i=1}^k \sum_{j=1}^k v_i a_{ij}^{[\ell]} v_j$$

$$\left(\mathrm{sum}_v(A) \right)^\ell \leq \left(v^T v \right)^{\ell-1} \mathrm{sum}_v(A^\ell) \;.$$

The special case where $v(x) = 1$ corresponds to the conjecture of Atkinson et al. The corresponding special case of Theorem 4.20 would assume that v is the all-ones vector ($v = \mathbf{1}_n$).

More information on related results can be found in the book of Mitrinović, Pečarić and Fink [MPF91, Ch. IX]. Integral inequalities (for probability measure spaces) corresponding to Theorem 7.6 were derived by Feng and Tonge [FT10].

9.2 Our Results

9.2.1 Hermitian Kernels

From Theorem 4.9, i.e., from

$$\text{sum}_v\left(A^{2a+c}\right)\cdot\text{sum}_v\left(A^{2a+2b+c}\right) \leq \text{sum}_v\left(A^{2a}\right)\cdot\text{sum}_v\left(A^{2(a+b+c)}\right) ,$$

we conclude the following results for Hermitian kernels.

Theorem 9.1
If $K(x,y) = \overline{K(y,x)}$ is a Hermitian kernel and $v(x) \in \mathbb{C}$, then

$$\left(\int_0^a \int_0^a \overline{v(x)}\,v(y)\,K_{2a+c}(x,y)\,dx\,dy\right)\left(\int_0^a \int_0^a \overline{v(x)}\,v(y)\,K_{2a+2b+c}(x,y)\,dx\,dy\right) \leq$$
$$\left(\int_0^a \int_0^a \overline{v(x)}\,v(y)\,K_{2a}(x,y)\,dx\,dy\right)\left(\int_0^a \int_0^a \overline{v(x)}\,v(y)\,K_{2(a+b+c)}(x,y)\,dx\,dy\right) .$$

Corollary 9.1
If $K(x,y) = \overline{K(y,x)}$ is a Hermitian kernel, then

$$\left(\int_0^a \int_0^a K_{2a+c}(x,y)\,dx\,dy\right)\left(\int_0^a \int_0^a K_{2a+2b+c}(x,y)\,dx\,dy\right) \leq$$
$$\left(\int_0^a \int_0^a K_{2a}(x,y)\,dx\,dy\right)\left(\int_0^a \int_0^a K_{2(a+b+c)}(x,y)\,dx\,dy\right) .$$

9.2.2 Nonnegative Symmetric Kernels

From Theorem 4.28, i.e., from

$$\left(\text{sum}_v\left(A^{2\ell+p}\right)\right)^k \leq \left(\text{sum}_v\left(A^{2\ell}\right)\right)^{k-1}\cdot\text{sum}_v\left(A^{2\ell+pk}\right) ,$$

we obtain the following results for nonnegative symmetric kernels.

Theorem 9.2
If $K(x,y) = K(y,x) \geq 0$ is a nonnegative symmetric kernel and $v(x) \in \mathbb{R}_{\geq 0}$, then

$$\left(\int_0^a \int_0^a v(x)\,v(y)\,K_{2\ell+p}(x,y)\,dx\,dy\right)^k \leq$$
$$\left(\int_0^a \int_0^a v(x)\,v(y)\,K_{2\ell}(x,y)\,dx\,dy\right)^{k-1}\int_0^a \int_0^a v(x)\,v(y)\,K_{2\ell+pk}(x,y)\,dx\,dy .$$

Corollary 9.2
If $K(x,y) = K(y,x) \geq 0$ is a nonnegative symmetric kernel, then

$$\left(\int_0^a \int_0^a K_{2\ell+p}(x,y)\,dx\,dy\right)^k \leq \left(\int_0^a \int_0^a K_{2\ell}(x,y)\,dx\,dy\right)^{k-1}\int_0^a \int_0^a K_{2\ell+pk}(x,y)\,dx\,dy .$$

9.2.3 Nonnegative Kernels

From Theorem 6.8, we conclude the following for nonsymmetric kernels.

Theorem 9.3
If $K(x,y) \geq 0$ is a nonnegative kernel and $u(x), v(x) \in \mathbb{R}_{\geq 0}$, then

$$\left(\int_0^a \int_0^b u(x)\,v(y)\,K(x,y)\,dx\,dy \right)^{2k+1} \leq \left(\frac{\int_0^a u(x)^2\,dx + \int_0^b v(y)^2\,dy}{2} \right)^{2k} \cdot$$
$$\int_0^a \int_0^b u(x)\,v(y)\,K_{(2k+1)}(x,y)\,dx\,dy \;,$$

$$\left(\int_0^a \int_0^b u(x)\,v(y)\,K(x,y)\,dx\,dy \right)^{2k} \leq \left(\frac{\int_0^a u(x)^2\,dx + \int_0^b v(y)^2\,dy}{2} \right)^{2k-1} \cdot$$
$$\frac{\int_0^a \int_0^a u(x)\,u(y)\,K_{(2k)}(x,y)\,dx\,dy + \int_0^b \int_0^b v(x)\,v(y)\,K^*_{(2k)}(x,y)\,dx\,dy}{2} \cdot$$

Here, $K^*_{(2k)}(x,y)$ denotes the analog of $(A^*A)^k$ in an equivalent way like $K_{(2k)}(x,y)$ represents the analog of $(AA^*)^k$.

Similar results can be deduced from the other theorems in the second and third part of this work (e.g., Theorems 4.8, 4.30, and 6.10), provided that suitable preconditions are fulfilled by the kernel function K and the scaling functions u and v.

Notice that these results can be extended to subsets of the domain of definition. Similar to the finite discrete case, where we used characteristic vectors, this can be accomplished by using the *characteristic function* of the subset (which has value 1 for all elements within the set and value 0 for all arguments that are not in the set).

9.3 Sidorenko's Conjecture

Sidorenko [Sid91; Sid92] made the following conjecture concerning integral inequalities related to undirected bipartite graphs. Suppose that H is an undirected graph with m_H edges. A *graph homomorphism* from a graph H to a graph G is a mapping $f : V(H) \mapsto V(G)$ such that, for each edge $\{u,v\}$ of H, $\{f(u), f(v)\}$ is an edge of G. Suppose that $h_H(G)$ denotes the number of homomorphisms from H to G.

There are several important examples. If H is a path consisting of k edges, then $h_H(G)$ denotes the number of k-step walks in G. If H is a cycle consisting of k edges, then $h_H(G)$ denotes the number of closed k-step walks in G. If H is a star consisting of k edges, then $h_H(G)$ denotes the sum of the k-th vertex degree powers of G. Other relations refer to the number of independent sets, the number of k-colorings, or bounds for the size of a maximum cut of G. For a survey on the topic, see Borgs et al. [BCL$^+$06].

Consider the fraction of mappings which are homomorphisms

$$t_H(G) := h_H(G)/|V(G)|^{|V(H)|} \ .$$

Then, Sidorenko conjectured that for every bipartite graph H and every graph G,

$$t_H(G) \geq t_{K_2}(G)^{m_H} \ .$$

This corresponds to

$$\frac{h_H(G)}{n_G^{n_H}} \geq \left(\frac{h_{K_2}(G)}{n_G^2}\right)^{m_H} = \left(\frac{2m_G}{n_G^2}\right)^{m_H} \ .$$

Consider measure spaces $\Omega = (X, \mu)$, $\Lambda = (Y, \nu)$ where μ and ν are finite σ-additive measures defined on a σ-algebra of subsets of the sets X and Y. For a measure space $\Omega = (X, \mu)$, let $K(\Omega)$ denote the class of nonnegative, bounded and measurable functions on X.

Conjecture 9.1 (Sidorenko)
For all bipartite graphs $G = (V, E) = (U \uplus W, E)$ with $U = \{u_1, \ldots, u_{n_1}\}$, $W = \{w_1, \ldots, w_{n_2}\}$, all measure spaces $\Omega = (X, \mu)$, $\Lambda = (Y, \nu)$, and any function $h \in K(\Omega \otimes \Lambda)$, the following inequality holds:

$$\int \prod_{(u_i, w_j) \in E} h(x_i, y_j) \, d\mu^{n_1} \, d\nu^{n_2} \geq \left(\int h(x, y) \, d\mu \, d\nu\right)^{|E|} d\mu(X)^{n_1 - |E|} \, d\nu(Y)^{n_2 - |E|} \ .$$

In particular, if both of the two measures correspond to the Lebesgue measure, Sidorenko [Sid93] conjectured the following.

Conjecture 9.2 (Sidorenko)
Suppose that μ denotes the Lebesgue measure on $[0, 1]$ and $h(x, y)$ is a bounded nonnegative function that is measurable on $[0, 1]^2$. For any bipartite graph $G = (V, E) = (U \uplus W, E)$ with $n_1 := |U|$, $n_2 := |W|$, $U = \{u_1, \ldots, u_{n_1}\}$, $W = \{w_1, \ldots, w_{n_2}\}$, and $E \subseteq U \times W$, we have

$$\int \prod_{(u_i, w_j) \in E} h(x_i, y_j) \, d\mu^{n_1 + n_2} \geq \left(\int h \, d\mu^2\right)^{|E|} \ .$$

By Theorems 4.20 and 4.22 (Mulholland and Smith [MS59] and Blakley and Roy [BR65]), the conjecture is valid if h is symmetric and G is a path. Theorem 6.7 by Pate [Pat12] implies that Sidorenko's conjecture is also true if G is a path and h is not necessarily symmetric. Hatami [Hat10] showed that the conjecture is true for certain graphs that include the hypercubes.

Sidorenko [Sid93] notes that integrals of the form

$$\int \prod_{(i, j) \in E} h(x_i, y_j) \, d\mu^{n + m}$$

are quite common. He remarks that they are called Mayer integrals in classical statistical mechanics, Feynman integrals in quantum field theory, and multicenter integrals in quantum chemistry [Ste67; DLM59].

We propose to use the more general approach of homomorphisms in *directed* graphs. This seems to be a more appropriate modeling, which can also be extended to functions of r variables and homomorphisms in r-uniform hypergraphs. In this sense, the conjecture is also true (by Theorem 6.7) if G is an alternating directed path of odd length (and h does not have to be symmetric). Using Theorem 6.5, Pate showed that Conjecture 9.2 is also true for nonsymmetric h if G is an alternating path of even length. In a sense, the theorem is also true for alternating walks of an even length, if the arithmetic mean of the two different types (starting forward or backward) is used.

Theorem 8.14 (see also Corollary 8.12) implies an inequality where the corresponding graphs are directed cycles. In the case of loop-free undirected graphs, such a directed cycle of length 2 corresponds to a single undirected edge, which yields another proof for the conjecture in the case of even cycles. Conlon, Fox and Sudakov [CFS10] proved the conjecture for all bipartite graphs containing a vertex that is connected to all vertices on the other side of the bipartition. Other results have been obtained more recently by Li and Szegedy [LS11], Szegedy [Sze15] and Kim, Lee and Lee [KLL16]. For further discussions, in particular on the connection to graphons, see Lovász [Lov09; Lov12].

Our Results.

Our results provide a possibility to bound the normalized number of homomorphisms not only by the edge density, but also by other homomorphism numbers. In particular, we are now able to give bounds between different path homomorphism numbers for paths of different lengths. The same applies to cycles.

Conclusion

> One can measure the importance of a scientific work
> by the number of earlier publications rendered superfluous by it.
>
> *David Hilbert*

In this work, we discussed many different inequalities for walks in graphs and for entry sums of matrix powers. It seems to be the first work that studies these relations in such a broad way. The result is a *systematic overview* of known results together with new and more general ones. Surely, these unifications and generalizations do not make superfluous any former work. Every single result is a step towards a more powerful statement, but without the specialized results, the more general ones would probably never exist.

It was one of the main goals to provide *simple arguments* for *explaining why* the different conceivable inequalities hold true or not. With a single exception, this goal has surely been attained. We should be aware that there are probably many possible ways to prove the inequalities mentioned in this book. The problem is how to find them, and the even bigger problem is to find the elementary ones. Once we have a simple proof, it is easy to reproduce it. So, if these results are easy to see now then it is one of the major achievements of our work. Now it is possible to explain most of these results and their connections to an average undergraduate student in natural sciences, mathematics, computer science, or engineering.

Bibliography

[AAA⁺03]   Ittai Abraham, Baruch Awerbuch, Yossi Azar, Yair Bartal, Dahlia Malkhi and Elan Pavlov. A generic scheme for building overlay networks in adversarial scenarios. In *Proceedings of the 17th International Parallel and Distributed Processing Symposium (IPDPS'03)*. IEEE, April 2003 (cited on page 32).

[AC07]   Reid Andersen and Sebastian M. Cioabă. Spectral densest subgraph and independence number of a graph. *Journal of Universal Computer Science*, 13(11):1501–1513, 2007 (cited on page 23).

[ACŠ12]   Vesna Andova, Nathann Cohen and Riste Škrekovski. A note on Zagreb indices inequality for trees and unicyclic graphs. *Ars Mathematica Contemporanea*, 5(1):73–76, 2012 (cited on pages 30, 92).

[ADRS14]   Hosam Abdo, Darko Dimitrov, Tamás Réti and Dragan Stevanović. Estimating the spectral radius of a graph by the second Zagreb index. *MATCH: Communications in Mathematical and in Computer Chemistry*, 72(3):741–751, 2014 (cited on page 152).

[AF02]   David Aldous and James Allen Fill. *Reversible Markov Chains and Random Walks on Graphs*. Unfinished monograph, 2002 (cited on page 26).

[ÁFNW09]   Bernardo M. Ábrego, Silvia Fernández-Merchant, Michael G. Neubauer and William Watkins. Sum of squares of degrees in a graph. *Journal of Inequalities in Pure and Applied Mathematics*, 10(3), 2009 (cited on page 70).

[AFWZ95]   Noga Alon, Uriel Feige, Avi Wigderson and David Zuckerman. Derandomized graph products. *Computational Complexity*, 5(1):60–75, March 1995 (cited on pages 22, 23).

[AHL02]   Noga Alon, Shlomo Hoory and Nathan Linial. The Moore bound for irregular graphs. *Graphs and Combinatorics*, 18(1):53–57, March 2002 (cited on page 59).

[Aig67]   Martin Aigner. On the linegraph of a directed graph. *Mathematische Zeitschrift*, 102(1):56–61, February 1967 (cited on pages 32, 34).

[AK78]   Rudolf Ahlswede and Gyula O. H. Katona. Graphs with maximal number of adjacent pairs of edges. *Acta Mathematica Academiae Scientiarum Hungaricae*, 32(1-2):97–120, March 1978 (cited on page 70).

[Alb97]   Michael O. Albertson. The irregularity of a graph. *Ars Combinatoria*, 46:219–225, August 1997 (cited on page 25).

[AWM60]   F. V. Atkinson, Geoffrey A. Watterson and Patrick A. P. Moran. A matrix inequality. *The Quarterly Journal of Mathematics (Oxford Second Series)*, 11:137–140, 1960 (cited on pages 28, 110, 159, 160).

[Bal92]   Alexandru T. Balaban. Using real numbers as vertex invariants for third-generation topological indexes. *Journal of Chemical Information and Computer Sciences*, 32(1):23–28, January 1992 (cited on page 30).

[Bap10]   Ravindra B. Bapat. *Graphs and Matrices, Universitext*. Springer / Hindustan Book Agency, 2010 (cited on page 38).

[Bar45]   Edward W. Barankin. Bounds for the characteristic roots of a matrix. *Bulletin of the American Mathematical Society*, 51(10):767–770, October 1945 (cited on pages 135, 136).

[BB83]   Edwin F. Beckenbach and Richard Bellman. *Inequalities*, volume 30 of *Ergebnisse der Mathematik und ihrer Grenzgebiete (Neue Folge)*. Springer, 1983. Fourth Printing (cited on page 18).

[BC09]   Richard A. Brualdi and Dragoš M. Cvetković. *A Combinatorial Approach to Matrix Theory and Its Applications*, volume 52 of *Discrete Mathematics and Its Applications*. Chapman & Hall / CRC Press, 2009 (cited on pages 8, 10, 11, 35, 38).

[BCL+06]   Christian Borgs, Jennifer Chayes, László Lovász, Vera T. Sós and Katalin Vesztergombi. Counting graph homomorphisms. In *Topics in Discrete Mathematics. Dedicated to Jarik Nešetřil on the Occasion of his 60th Birthday*. Volume 26, Algorithms and Combinatorics, pages 315–371. Springer, 2006 (cited on page 162).

[BD66]   George R. Blakley and Robert D. Dixon. Hölder type inequalities in cones. *Journal of Mathematical Analysis and Applications*, 14(1):1–4, April 1966 (cited on pages 58, 59, 62, 63).

[BD84]   Robert C. Brigham and Ronald D. Dutton. Bounds on graph spectra. *Journal of Combinatorial Theory, Series B*, 37(3):228–234, December 1984 (cited on page 151).

[Bee69]   Paul R. Beesack. Inequalities involving iterated kernels and convolutions. *Univ. Beograd, Publ. Elektrotehn. Fak., Ser. Mat. Fiz.*, 274–301(276):11–16, 1969 (cited on page 160).

[Bei68]   Lowell W. Beineke. On derived graphs and digraphs. In *Beiträge zur Graphentheorie*, pages 17–23. B. G. Teubner, Leipzig, 1968 (cited on page 31).

[BEK13]   Michele Benzi, Ernesto Estrada and Christine Klymko. Ranking hubs and authorities using matrix functions. *Linear Algebra and its Applications*, 438(5):2447–2474, March 2013 (cited on page 34).

[Bel92]   Francis K. Bell. A note on the irregularity of graphs. *Linear Algebra and its Applications*, 161:45–54, January 1992 (cited on page 25).

[Ben14]   Michele Benzi. A note on walk entropies in graphs. *Linear Algebra and its Applications*, 445:395–399, March 2014 (cited on page 34).

[Ber09]   Dennis S. Bernstein. *Matrix Mathematics. Theory, Facts, and Formulas*. Princeton University Press, 2nd edition, 2009 (cited on page 81).

[BF94]   Jean-Claude Bermond and Pierre Fraigniaud. Broadcasting and gossiping in de Bruijn networks. *SIAM Journal on Computing*, 23(1):212–225, February 1994 (cited on page 32).

[BH85]   Richard A. Brualdi and Alan J. Hoffman. On the spectral radius of $(0, 1)$-matrices. *Linear Algebra and its Applications*, 65:133–146, February 1985 (cited on pages 147, 151).

[BHNR03]   William Banks, Asma Harcharras, Stefan Neuwirth and Eric Ricard. Matrix inequalities with applications to the theory of iterated kernels. *Linear Algebra and its Applications*, 362:275–286, March 2003 (cited on pages 109–111).

[BHV94]   Winfried Bruns, Jürgen Herzog and Udo Vetter. Syzygies and walks. In A. Simis, N. V. Trung and G. Valla, editors, *Commutative Algebra (Proceedings, Trieste 1992)*, pages 36–57. World Scientific, 1994 (cited on page 21).

[Big74]   Norman Biggs. *Algebraic Graph Theory*, volume 67 of *Cambridge Tracts in Mathematics*. Cambridge University Press, 1974 (cited on page 38).

[Big93]   Norman Biggs. *Algebraic Graph Theory, Cambridge Mathematical Library*. Cambridge University Press, 2nd edition, 1993 (cited on page 38).

[BK11]   Hoda Bidkhori and Shaunak Kishore. A bijective proof of a theorem of Knuth. *Combinatorics, Probability and Computing*, 20(1):11–25, January 2011 (cited on pages 31, 32).

[BK13]   Michele Benzi and Christine Klymko. Total communicability as a centrality measure. *Journal of Complex Networks*, 1(2):124–149, December 2013 (cited on page 34).

[BK15]   Michele Benzi and Christine Klymko. On the limiting behavior of parameter-dependent network centrality measures. *SIAM Journal on Matrix Analysis and Applications*, 36(2):686–706, 2015 (cited on page 34).

[BK89]   Andrei Z. Broder and Anna R. Karlin. Bounds on the cover time. *Journal of Theoretical Probability*, 2(1):101–120, January 1989 (cited on page 26).

[BKMR05]   Joachim Braun, Adalbert Kerber, Markus Meringer and Christoph Rücker. Similarity of molecular descriptors: the equivalence of Zagreb indices and walk counts. *MATCH: Communications in Mathematical and in Computer Chemistry*, 54(1):163–176, 2005 (cited on page 29).

[Blu91]   Avrim Blum. *Algorithms for Approximate Graph Coloring*. PhD thesis, MIT Laboratory for Computer Science, 1991 (cited on page 23).

[BP07]   Jean Berstel and Dominique Perrin. The origins of combinatorics on words. *European Journal of Combinatorics*, 28(3):996–1022, April 2007 (cited on page 21).

[BP89]   Jean-Claude Bermond and Claudine Peyrat. De Bruijn and Kautz networks: a competitor for the hypercube? In *Proceedings of the 1st European Workshop on Hypercubes and Distributed Computers (Rennes, France)*, pages 279–293. North Holland, 1989 (cited on page 32).

[BP94]   Abraham Berman and Robert J. Plemmons. *Nonnegative Matrices in the Mathematical Sciences*, volume 9 of *Classics in Applied Mathematics*. SIAM, 1994 (cited on pages 27, 123).

[BR65]   George R. Blakley and Prabir Roy. A Hölder type inequality for symmetric matrices with nonnegative entries. *Proceedings of the American Mathematical Society*, 16(6):1244–1245, December 1965 (cited on pages 22, 58, 61, 111, 114, 163).

[BR97]   Ravindra B. Bapat and Thirukkannamangai E. S. Raghavan. *Nonnegative Matrices and Applications*, volume 64 of *Encyclopedia of Mathematics and its Applications*. Cambridge University Press, 1997 (cited on pages 123, 135).

[Bru10]   Richard A. Brualdi. Spectra of digraphs. *Linear Algebra and its Applications*, 432(9):2181–2213, April 2010 (cited on page 149).

[Bru11]   Richard A. Brualdi. *The Mutually Beneficial Relationship of Graphs and Matrices*, number 115 of *CBMS Regional Conference Series in Mathematics*. American Mathematical Society, 2011 (cited on pages 38, 109, 135).

[BS13]   Brian K. Butler and Paul H. Siegel. Sharp bounds on the spectral radius of nonnegative matrices and digraphs. *Linear Algebra and its Applications*, 439(5):1468–1478, September 2013 (cited on pages 145, 148).

[BS92]   Piotr Berman and Georg Schnitger. On the complexity of approximating the independent set problem. *Information and Computation*, 96(1):77–94, January 1992 (cited on page 23).

[BT12]   Béla Bollobás and Mykhaylo Tyomkyn. Walks and paths in trees. *Journal of Graph Theory*, 70(1):54–66, May 2012 (cited on page 24).

[Buc33]   Aleksandr A. Buchstab. Об одной метрической задаче аддитивной теории чисел (On a metric problem of additive number theory). *Математический сборник (Matematiceskij Sbornik)*, 40:190–195, 1933 (cited on pages 34, 123).

[Bul03]   Peter S. Bullen. *Handbook of Means and Their Inequalities*, volume 560 of *Mathematics and Its Applications*. Kluwer Academic Publishers, 2nd edition, 2003 (cited on pages 14, 18).

[Bun59]   Viktor Yakovlevich Bunyakovsky. Sur quelques inégalités concernant les intégrales ordinaires et les intégrales aux différences finies. *Mé-moires de l'Académie Impériale des sciences de St.-Pétersbourg, VIIe Série*, I(9):1–18, 1859 (cited on page 13).

[Bür11]   Reinhard Bürger. Some mathematical models in evolutionary genetics. In Fabio A. C. C. Chalub and José Francisco Rodrigues, editors, *The Mathematics of Darwin's Legacy*, Mathematics and Biosciences in Interaction, pages 67–89. Birkhäuser / Springer, 2011 (cited on page 28).

[But06a]   Steven Butler. Relating singular values and discrepancy of weighted directed graphs. In *Proceedings of the 17th ACM-SIAM Symposium on Discrete Algorithms (SODA'06)*, pages 1112–1116, January 2006 (cited on pages 8, 112).

[But06b]   Steven Butler. Using discrepancy to control singular values for nonnegative matrices. *Linear Algebra and its Applications*, 419(2–3):486–493, December 2006 (cited on page 112).

[But08]   Steven Butler. *Eigenvalues and Structures of Graphs*. PhD thesis, University of California, San Diego, 2008 (cited on page 112).

[But13]   Brian K. Butler. *Error Floors of LDPC Codes and Related Topics*. PhD thesis, University of California, San Diego, 2013 (cited on pages 145, 148).

[BW05]   Lowell W. Beineke and Robin J. Wilson, editors. *Topics in Algebraic Graph Theory*, volume 102 of *Encyclopedia of Mathematics and its Applications*. Cambridge University Press, 2005 (cited on page 38).

[BZ01]   Abraham Berman and Xiao-Dong Zhang. On the spectral radius of graphs with cut vertices. *Journal of Combinatorial Theory, Series B*, 83(2):233–240, November 2001 (cited on page 149).

[Car72]   Bille C. Carlson. The logarithmic mean. *The American Mathematical Monthly*, 79(6):615–618, 1972 (cited on page 82).

[Cau21]   Augustin-Louis Cauchy. *Cours d'Analyse de l'École Royale Polytechnique ; Première Partie. Analyse algébrique*. Debure frères, Paris, 1821 (cited on page 13).

[CCR14]   Gilles Caporossi, Dragoš M. Cvetković and Peter Rowlinson. Spectral reconstruction and isomorphism of graphs using variable neighbourhood search. *Bulletin de l'Académie serbe des sciences et des arts, Classe des Sciences mathématiques et naturelles, Sciences mathématiques*, CXLVI(39):23–38, 2014 (cited on page 34).

[CDGT88]   Dragoš M. Cvetković, Michael Doob, Ivan Gutman and Aleksandar Torgašev. *Recent Results in the Theory of Graph Spectra*, volume 36 of *Annals of Discrete Mathematics*. North-Holland, 1988 (cited on pages 35, 38).

[CDS14]   Shujuan Cao, Matthias Dehmer and Yongtang Shi. Extremality of degree-based graph entropies. *Information Sciences*, 278:22–33, September 2014 (cited on pages 34, 71).

[CDS79]   Dragoš M. Cvetković, Michael Doob and Horst Sachs. *Spectra of Graphs – Theory and Applications.* Deutscher Verlag der Wissenschaften Berlin, 1979 (cited on pages 20, 38, 41, 43, 146).

[CFS10]   David Conlon, Jacob Fox and Benny Sudakov. An approximate version of Sidorenko's conjecture. *Geometric and Functional Analysis*, 20(6):1354–1366, December 2010 (cited on page 164).

[CG09]   Dragoš M. Cvetković and Ivan Gutman, editors. *Applications of Graph Spectra*, volume 13(21) of *Zbornik radova*. Mathematical Institute SANU, Belgrade, June 2009 (cited on page 27).

[CG11]   Dragoš M. Cvetković and Ivan Gutman, editors. *Selected Topics on Applications of Graph Spectra*, volume 14(22) of *Zbornik radova*. Mathematical Institute SANU, Belgrade, January 2011 (cited on page 27).

[CG66]   Dorwin Cartwright and Terry C. Gleason. The number of paths and cycles in a digraph. *Psychometrika*, 31(2):179–199, June 1966 (cited on page 33).

[Cha12]   Françoise Chatelin. *Eigenvalues of Matrices*, volume 71 of *Classics in Applied Mathematics*. SIAM, revised edition edition, 2012 (cited on page 135).

[Che83]   Pafnuty L. Chebyshev. Объ одномъ рядѣ, доставляющемъ предѣльныя величины интеграловъ при разложении подъ-интегральной функции на множители. *Записки Императорской Академии наук (Санкт-Петербург)*, XLVII(4), 1883 (cited on page 15).

[Chu97]   Fan R. K. Chung. *Spectral Graph Theory*, number 92 of *CBMS Regional Conference Series in Mathematics*. American Mathematical Society, 1997 (cited on page 38).

[Cio06]   Sebastian M. Cioabă. Sums of powers of the degrees of a graph. *Discrete Mathematics*, 306(16):1959–1964, August 2006 (cited on page 71).

[Cio11]   Sebastian M. Cioabă. Some applications of eigenvalues of graphs. In Matthias Dehmer, editor, *Structural Analysis of Complex Networks*, part 14, pages 357–379. Birkhäuser, 2011 (cited on page 27).

[CL14]   Péter Csikvári and Zhicong Lin. Graph homomorphisms between trees. *The Electronic Journal of Combinatorics*, 21(4):P4.9/1–38, October 2014 (cited on page 24).

[CR90]   Dragoš M. Cvetković and Peter Rowlinson. The largest eigenvalue of a graph: a survey. *Linear and Multilinear Algebra*, 28(1):3–33, October 1990 (cited on page 153).

[CRS09]   Dragoš M. Cvetković, Peter Rowlinson and Slobodan K. Simić. *An Introduction to the Theory of Graph Spectra*, volume 75 of *London Mathematical Society Student Texts*. Cambridge University Press, 2009 (cited on page 38).

[CRS97]   Dragoš M. Cvetković, Peter Rowlinson and Slobodan K. Simić. *Eigenspaces of Graphs*, volume 66 of *Encyclopedia of Mathematics and its Applications*. Cambridge University Press, 1997 (cited on page 38).

[CS57]   Lothar Collatz and Ulrich Sinogowitz. Spektren endlicher Grafen. *Abhandlungen aus dem Mathematischen Seminar der Universität Hamburg*, 21(1):63–77, December 1957 (cited on pages 25, 37, 151).

[CS66]   Gary Chartrand and M. James Stewart. Total digraphs. *Canadian Mathematical Bulletin*, 9:171–176, 1966 (cited on page 31).

[Csi10]   Péter Csikvári. On a poset of trees. *Combinatorica*, 30(2):125–137, March 2010 (cited on page 24).

[Cve69]   Dragoš M. Cvetković. Spectrum of the graph of n-tuples. *Univ. Beograd, Publ. Elektrotehn. Fak., Ser. Mat. Fiz.*, 274–301(288):91–95, 1969 (cited on page 41).

[Cve70a]   Dragoš M. Cvetković. A note on paths in the p-sum of graphs. *Univ. Beograd, Publ. Elektrotehn. Fak., Ser. Mat. Fiz.*, 302–319(311):49–51, 1970 (cited on page 41).

[Cve70b]   Dragoš M. Cvetković. Die Zahl der Wege eines Grafen. *Glasnik Matematički Serija III*, 5(2):205–210, 1970 (cited on pages 20, 41).

[Cve70c]   Dragoš M. Cvetković. The generating function for variations with restrictions and paths of the graph and self-complementary graphs. *Univ. Beograd, Publ. Elektrotehn. Fak., Ser. Mat. Fiz.*, 320–328(322):27–34, 1970 (cited on pages 20, 38, 41).

[Cve71]   Dragoš M. Cvetković. Graphs and their spectra. *Univ. Beograd, Publ. Elektrotehn. Fak., Ser. Mat. Fiz.*, 354–456(354):1–50, 1971 (cited on pages 38, 41, 43).

[Cve73]   Dragoš M. Cvetković. On a graph theory problem of M. Koman. *Časopis pro pěstování matematiky*, 98(3):233–236, 1973 (cited on page 41).

[CWW+08]   Deepayan Chakrabarti, Yang Wang, Chenxi Wang, Jurij Leskovec and Christos Faloutsos. Epidemic thresholds in real networks. *ACM Transactions on Information and System Security*, 10(4):13:1–13:26, January 2008 (cited on page 27).

[Das03]   Kinkar Ch. Das. Sharp bounds for the sum of the squares of the degrees of a graph. *Kragujevac Journal of Mathematics*, 25:31–49, 2003 (cited on page 70).

[Das04]   Kinkar Ch. Das. Maximizing the sum of the squares of the degrees of a graph. *Discrete Mathematics*, 285(1-3):57–66, August 2004 (cited on pages 30, 70).

[DB08]   Kinkar Ch. Das and Ravindra B. Bapat. A sharp upper bound on the spectral radius of weighted graphs. *Discrete Mathematics*, 308(15):3180–3186, August 2008 (cited on page 138).

[dBru46]   Nicolaas G. de Bruijn. A combinatorial problem. *Koninklijke Nederlandsche Akademie van Wetenschappen – Proceedings of the Section of Sciences*, 49(7):758–764, 1946 (cited on pages 31, 32).

[dBru75]   Nicolaas G. de Bruijn. Acknowledgement of priority to C. Flye Sainte-Marie on the counting of circular arrangements of 2^n zeros and ones that show each n-letter word exactly once. T.H.-Report 75-WSK-06, Technological University Eindhoven, The Netherlands, Department of Mathematics, 1975 (cited on page 32).

[dCae98]   Dominique de Caen. An upper bound on the sum of squares of degrees in a graph. *Discrete Mathematics*, 185:245–248, April 1998 (cited on pages 30, 70).

[Dea73]   Michael A. B. Deakin. On the status of mean fitness maximization as a governing principle in evolution. *Bulletin of Mathematical Biology*, 35:43–50, 1973 (cited on page 28).

[dFBRS14]   Maria Aguieiras A. de Freitas, Andréa Soares Bonifácio, María Robbiano and Bernardo San Martín. On matrices associated to directed graphs and applications. *Linear Algebra and its Applications*, 442:156–164, February 2014 (cited on pages 31, 32).

[DFG$^+$11]   Tomislav Došlić, Boris Furtula, Ante Graovac, Ivan Gutman, Sirous Moradi and Zahra Yaramahdi. On vertex-degree-based molecular structure descriptors. *MATCH: Communications in Mathematical and in Computer Chemistry*, 66(2):613–626, 2011 (cited on page 30).

[DG03a]   Andreas Dress and Ivan Gutman. Asymptotic results regarding the number of walks in a graph. *Applied Mathematics Letters*, 16(3):389–393, April 2003 (cited on page 48).

[DG03b]   Andreas Dress and Ivan Gutman. The number of walks in a graph. *Applied Mathematics Letters*, 16(5):797–801, July 2003 (cited on pages 48, 50, 53, 64, 65, 109).

[Diu94]   Mircea V. Diudea. Molecular topology. 16. Layer matrices in molecular graphs. *Journal of Chemical Information and Computer Sciences*, 34(5):1064–1071, September 1994 (cited on page 30).

[Diu96]   Mircea V. Diudea. Walk numbers ^eW_M: Wiener-type numbers of higher rank. *Journal of Chemical Information and Computer Sciences*, 36(3):535–540, May 1996 (cited on page 30).

[DLH93]   Ding-Zhu Du, Yuh-Dauh Lyuu and D. Frank Hsu. Line digraph iterations and connectivity analysis of de Bruijn and Kautz graphs. *IEEE Transactions on Computers*, 42(5):612–616, May 1993 (cited on page 33).

[DLM59]   Raymond Daudel, Roland Lefebvre and Carl Moser. *Quantum Chemistry. Methods and Applications*. Wiley, 1959 (cited on page 164).

[DLS15]   Matthias Dehmer, Xueliang Li and Yongtang Shi. Connections between generalized graph entropies and graph energy. *Complexity*, 21(1):35–41, 2015 (cited on page 34).

[DM15]   Matthias Dehmer and Abbe Mowshowitz. A case study of cracks in the scientific enterprise: reinvention of information-theoretic measures for graphs. *Complexity*, 2015. to appear (cited on page 34).

[DMB91]   Mircea V. Diudea, Ovidiu Minailiuc and Alexandru T. Balaban. Molecular topology. IV. Regressive vertex degrees (new graph invariants) and derived topological indices. *Journal of Computational Chemistry*, 12(5):527–535, June 1991 (cited on page 30).

[dOOJdA13]   Joelma Ananias de Oliveira, Carla Silva Oliveira, Claudia Justel and Nair Maria Maia de Abreu. Measures of irregularity of graphs. *Pesquisa Operacional*, 33(3):383–393, December 2013 (cited on page 25).

[DTG94]   Mircea V. Diudea, Mihai Topan and Ante Graovac. Molecular topology. 17. Layer matrices of walk degrees. *Journal of Chemical Information and Computer Sciences*, 34(5):1072–1078, September 1994 (cited on page 30).

[DW83]   Emeric Deutsch and Helmut Wielandt. Nested bounds for the Perron root of a nonnegative matrix. *Linear Algebra and its Applications*, 52/53:235–251, July 1983 (cited on page 142).

[EdlPH14]   Ernesto Estrada, José A. de la Peña and Naomichi Hatano. Walk entropies in graphs. *Linear Algebra and its Applications*, 443:235–244, February 2014 (cited on page 34).

[Edw00]   Anthony W. F. Edwards. *Foundations of Mathematical Genetics*. Cambridge University Press, 2nd edition, 2000 (cited on page 28).

[Edw67]   Anthony W. F. Edwards. Fundamental theorem of natural selection. *Nature*, 215(5100):537–538, July 1967 (cited on page 28).

[Edw77]   C. S. Edwards. The largest vertex degree sum for a triangle in a graph. *Bulletin of the London Mathematical Society*, 9(2):203–208, July 1977 (cited on page 25).

[Edw94]   Anthony W. F. Edwards. The fundamental theorem of natural selection. *Biological Reviews*, 69(4):443–474, November 1994 (cited on page 28).

[EE83]   C. S. Edwards and Clive Elphick. Lower bounds for the clique and the chromatic numbers of a graph. *Discrete Applied Mathematics*, 5(1):51–64, January 1983 (cited on page 152).

[EGG98]   Ernesto Estrada, Nicolais Guevara and Ivan Gutman. Extension of edge connectivity index. Relationships to line graph indices and QSPR applications. *Journal of Chemical Information and Computer Sciences*, 38(3):428–431, May 1998 (cited on page 30).

[EH85]   Abdol-Hossein Esfahanian and S. Louis Hakimi. Fault-tolerant routing in deBruijn communication networks. *IEEE Transactions on Computers*, C-34(9):777–788, September 1985 (cited on page 32).

[EIG12]   Mehdi Eliasi, Ali Iranmanesh and Ivan Gutman. Multiplicative versions of first Zagreb index. *MATCH: Communications in Mathematical and in Computer Chemistry*, 68(1):217–230, 2012 (cited on page 83).

[ER96]   Ernesto Estrada and Alain Ramírez. Edge adjacency relationships and molecular topographic descriptors. Definition and QSAR applications. *Journal of Chemical Information and Computer Sciences*, 36(4):837–843, July 1996 (cited on page 30).

[ES11]   Robert Elsässer and Thomas Sauerwald. Tight bounds for the cover time of multiple random walks. *Theoretical Computer Science*, 412(24):2623–2641, May 2011 (cited on page 26).

[ES46]   Paul Erdős and Arthur H. Stone. On the structure of linear graphs. *Bulletin of the American Mathematical Society*, 52(12):1087–1091, December 1946 (cited on page 23).

[ES66]   Paul Erdős and Miklós Simonovits. A limit theorem in graph theory. *Studia Scientiarum Mathematicarum Hungarica*, 1:51–57, 1966 (cited on page 23).

[ES82]   Paul Erdős and Miklós Simonovits. Compactness results in extremal graph theory. *Combinatorica*, 2(3):275–288, September 1982 (cited on pages 22–24, 58–60, 64, 68).

[Est09]   Ernesto Estrada. Spectral theory of networks: from biomolecular to ecological systems. In Matthias Dehmer and Frank Emmert-Streib, editors, *Analysis of Complex Networks*, pages 55–83. Wiley-VCH, 2009 (cited on page 27).

[Est96]   Ernesto Estrada. Spectral moments of the edge adjacency matrix in molecular graphs. 1. Definition and applications to the prediction of physical properties of alkanes. *Journal of Chemical Information and Computer Sciences*, 36(4):844–849, July 1996 (cited on page 30).

[Est97]   Ernesto Estrada. Spectral moments of the edge-adjacency matrix of molecular graphs. 2. Molecules containing heteroatoms and QSAR applications. *Journal of Chemical Information and Computer Sciences*, 37(2):320–328, March 1997 (cited on page 30).

[Est98]   Ernesto Estrada. Spectral moments of the edge adjacency matrix in molecular graphs. 3. Molecules containing cycles. *Journal of Chemical Information and Computer Sciences*, 38(1):23–27, January 1998 (cited on page 30).

[EV13]   Mehdi Eliasi and Damir Vukičević. Comparing the multiplicative Zagreb indices. *MATCH: Communications in Mathematical and in Computer Chemistry*, 69(3):765–773, 2013 (cited on pages 80, 83).

[EW10]   Warren J. Ewens and Geoffrey A. Watterson. Kingman and mathematical population genetics. In *Probability and Mathematical Genetics (Papers in Honour of Sir John Kingman)*, volume 378 of *London Mathematical Society Lecture Note Series*, pages 238–263. Cambridge University Press, 2010 (cited on page 28).

[EW14]   Clive Elphick and Pawel Wocjan. New measures of graph irregularity. *Electronic Journal of Graph Theory and Applications*, 2(1):52–65, 2014 (cited on page 25).

[FAYF85]   Miquel Ángel Fiol, Ignacio Alegre, J. Luis A. Yebra and Josep Fàbrega. Digraphs with walks of equal length between vertices. In *Proceedings of the 5th Int. Conference on the Theory and Applications of Graphs with Special Emphasis on Algorithms And Computer Science Applications (June 4-8, 1984, Kalamazoo, Michigan)*, pages 313–322, 1985 (cited on page 31).

[Fei95a]   Uriel Feige. A tight lower bound on the cover time for random walks on graphs. *Random Structures & Algorithms*, 6(4):433–438, July 1995 (cited on page 26).

[Fei95b]   Uriel Feige. A tight upper bound on the cover time for random walks on graphs. *Random Structures & Algorithms*, 6(1):51–54, January 1995 (cited on page 26).

[FG03]   Pierre Fraigniaud and Philippe Gauron. Brief announcement: an overview of the content-addressable network D2B. In *Proceedings of the 22nd ACM Symposium on Principles of Distributed Computing (PODC'03)*, page 151. ACM, July 2003 (cited on page 32).

[FG06]   Pierre Fraigniaud and Philippe Gauron. D2B: a de Bruijn based content-addressable network. *Theoretical Computer Science*, 355(1):65–79, April 2006 (cited on page 32).

[FG09]   Miquel Ángel Fiol and Ernest Garriga. Number of walks and degree powers in a graph. *Discrete Mathematics*, 309(8):2613–2614, April 2009 (cited on pages 71, 72, 127).

[FGE14]   Boris Furtula, Ivan Gutman and Süleyman Ediz. On difference of Zagreb indices. *Discrete Applied Mathematics*, 178:83–88, December 2014 (cited on page 30).

[FGN13]   Gholam Hossein Fath-Tabar, Ivan Gutman and Ramin Nasiri. Extremely irregular trees. *Bulletin de l'Académie serbe des sciences et des arts, Classe des Sciences mathématiques et naturelles, Sciences mathématiques*, CXLV(38):1–8, 2013 (cited on page 25).

[FK06]   Zoltán Füredi and André Kündgen. Moments of graphs in monotone families. *Journal of Graph Theory*, 51(1):37–48, January 2006 (cited on page 71).

[FKP01]   Uriel Feige, Guy Kortsarz and David Peleg. The dense k-subgraph problem. *Algorithmica*, 29(3):410–421, March 2001 (cited on page 22).

[FL92]   Miquel Ángel Fiol and Anna S. Lladó. The partial line digraph technique in the design of large interconnection networks. *IEEE Transactions on Computers*, 41(7):848–857, July 1992 (cited on pages 31, 32).

[Fly94]   Camille Flye Sainte-Marie. (Réponse) 48. *L'Intermédiaire des Mathématiciens*, I:107–110, 1894 (cited on page 32).

[FMS93]   Odile Favaron, Maryvonne Mahéo and Jean-François Saclé. Some eigenvalue properties in graphs (conjectures of Graffiti – II). *Discrete Mathematics*, 111(1-3):197–220, February 1993 (cited on pages 151, 156).

[Fri85]   Shmuel Friedland. The maximal eigenvalue of 0–1 matrices with pre-scribed number of ones. *Linear Algebra and its Applications*, 69:33–69, August 1985 (cited on pages 147, 151).

[Fri88]   Shmuel Friedland. Bounds on the spectral radius of graphs with e edges. *Linear Algebra and its Applications*, 101:81–86, April 1988 (cited on page 151).

[Fro08]   Georg Frobenius. Über Matrizen aus positiven Elementen. *Sitzungsberichte der Königlich Preussischen Akademie der Wissenschaften*:471–476, 1908 (cited on page 142).

[Fro09]   Georg Frobenius. Über Matrizen aus positiven Elementen. II. *Sitzungsberichte der Königlich Preussischen Akademie der Wissenschaften*:514–518, 1909 (cited on page 142).

[Fro12]   Georg Frobenius. Über Matrizen aus nicht negativen Elementen. *Sitzungsberichte der Königlich Preussischen Akademie der Wissenschaften*:456–477, 1912 (cited on page 142).

[FS09]   Philippe Flajolet and Robert Sedgewick. *Analytic Combinatorics*. Cambridge University Press, 2009 (cited on pages 6, 21).

[FS97]   Uriel Feige and Michael Seltser. On the densest k-subgraph problem. Technical report, Department of Applied Mathematics and Computer Science, Weizmann Institute Rehovot, Israel, September 1997 (cited on page 22).

[FT10]   Bao Qi Feng and Andrew Tonge. A Cauchy-Khinchin integral inequality. *Linear Algebra and its Applications*, 433(5):1024–1030, October 2010 (cited on pages 125, 160).

[FYA84]   Miquel Ángel Fiol, J. Luis A. Yebra and Ignacio Alegre. Line digraph iterations and the (d, k) digraph problem. *IEEE Transactions on Computers*, C-33(5):400–403, May 1984 (cited on page 33).

[Gan60]   Felix R. Gantmacher. *The Theory of Matrices*. Chelsea Publishing Company, 1960 (cited on pages 12, 40, 42, 142).

[Gau53a]   Werner Gautschi. The asymptotic behaviour of powers of matrices. *Duke Mathematical Journal*, 20(1):127–140, March 1953 (cited on page 138).

[Gau53b]   Werner Gautschi. The asymptotic behaviour of powers of matrices. II. *Duke Mathematical Journal*, 20(3):375–379, September 1953 (cited on page 138).

[GD04]   Ivan Gutman and Kinkar Ch. Das. The first Zagreb index 30 years after. *MATCH: Communications in Mathematical and in Computer Chemistry*, 50:83–92, February 2004 (cited on page 30).

[GE96]   Ivan Gutman and Ernesto Estrada. Topological indices based on the line graph of the molecular graph. *Journal of Chemical Information and Computer Sciences*, 36(3):541–543, May 1996 (cited on page 30).

[GFE14]   Ivan Gutman, Boris Furtula and Clive Elphick. Three new/old vertex-degree-based topological indices. *MATCH: Communications in Mathematical and in Computer Chemistry*, 72(3):617–632, 2014 (cited on pages 25, 30).

[GH68]   Dennis P. Geller and Frank Harary. Arrow diagrams are line di-
            graphs. *SIAM Journal on Applied Mathematics*, 16(6):1141–1145,
            November 1968 (cited on page 31).

[GHM77]   Christopher D. Godsil, Derek A. Holton and Brendan D. McKay.
            The spectrum of a graph. In *Combinatorial Mathematics V – Pro-
            ceedings of the 5th Australian Conference, Held at the Royal Mel-
            bourne Institute of Technology, August 24–26, 1976*, volume 622 of
            Lecture Notes in Mathematics, pages 91–117, 1977 (cited on page 38).

[GJ74]   F. Göbel and A. A. Jagers. Random walks on graphs. *Stochastic Pro-
            cesses and their Applications*, 2(4):311–336, 1974 (cited on page 26).

[GLZ09]   Ivan Gutman, Xueliang Li and Jianbin Zhang. Graph energy. In
            Matthias Dehmer and Frank Emmert-Streib, editors, *Analysis of
            Complex Networks*, pages 145–174. Wiley-VCH, 2009 (cited on
            page 155).

[GMT05]   Ayalvadi Ganesh, Laurent Massoulié and Donald Towsley. The ef-
            fect of network topology on the spread of epidemics. In *Proceedings of
            the 24th Annual Joint Conference of the IEEE Computer and Com-
            munications Societies (INFOCOM'05)*, volume 2, pages 1455–1466,
            March 2005 (cited on page 27).

[God93]   Christopher D. Godsil. *Algebraic Combinatorics*. Chapman & Hall,
            1993 (cited on page 38).

[Gol14a]   Felix Goldberg. New results on eigenvalues and degree deviation,
            March 2014. arXiv: `1403.2629v1` [`math.CO`] (cited on page 25).

[Gol14b]   Felix Goldberg. Spectral radius minus average degree: a better
            bound, July 2014. arXiv: `1407.4285v1` [`math.CO`] (cited on page 25).

[Gol15]   Felix Goldberg. A spectral bound for graph irregularity. *Czechoslovak
            Mathematical Journal*, 65(2):375–379, June 2015 (cited on page 25).

[Goo46]   Irving John Good. Normal recurring decimals. *Journal of the London
            Mathematical Society*, 21(3):167–169, July 1946 (cited on page 32).

[GR01]   Christopher D. Godsil and Gordon F. Royle. *Algebraic Graph The-
            ory*, volume 207 of *Graduate Texts in Mathematics*. Springer, 2001
            (cited on page 38).

[GR10]   Elías Gudiño and Juan Rada. A lower bound for the spectral radius
            of a digraph. *Linear Algebra and its Applications*, 433(1):233–240,
            July 2010 (cited on page 147).

[GR14]   Ivan Gutman and Tamás Réti. Zagreb group indices and beyond.
            International Journal of Chemical Modeling, 6(2-3):191–200, 2014
            (cited on page 30).

[Gra87]   Alexander Graham. *Nonnegative Matrices and Applicable Topics in
            Linear Algebra, Mathematics and its Applications*. Ellis Horwood
            Ltd, Chichester, 1987 (cited on pages 27, 123).

[GRR01]   Ivan Gutman, Christoph Rücker and Gerta Rücker. On walks in
            molecular graphs. *Journal of Chemical Information and Computer
            Sciences*, 41(3):739–745, May 2001 (cited on pages 29, 30, 48).

[GRTW75]   Ivan Gutman, Branko Ruščić, Nenad Trinajstić and Charles F. Wilcox Jr. Graph theory and molecular orbitals. XII. Acyclic polyenes. *The Journal of Chemical Physics*, 62(9):3399–3405, May 1975 (cited on page 29).

[Grü69]   Branko Grünbaum. Incidence patterns of graphs and complexes. In *The Many Facets of Graph Theory – Proceedings of the Conference held at Western Michigan University, Kalamazoo / MI, October 31 – November 2, 1968*, volume 110 of *Lecture Notes in Mathematics*, pages 115–128, 1969 (cited on page 31).

[GS64]   M. Gordon and G. R. Scantlebury. Non-random polycondensation: statistical theory of the substitution effect. *Transactions of the Faraday Socicty*, 60:604–621, 1964 (cited on page 29).

[GT72]   Ivan Gutman and Nenad Trinajstić. Graph theory and molecular orbitals. Total π-electron energy of alternant hydrocarbons. *Chemical Physics Letters*, 17(4):535–538, December 1972 (cited on pages 29, 155).

[GT73]   Ivan Gutman and Nenad Trinajstić. Graph theory and molecular orbitals. The loop rule. *Chemical Physics Letters*, 20(3):257–260, June 1973 (cited on page 155).

[Gut01]   Ivan Gutman. The energy of a graph: old and new results. In *Proceedings of the Euroconference Algebraic Combinatorics and Applications (ALCOMA'99)*, pages 196–211. Springer, 2001 (cited on page 155).

[Gut03]   Ivan Gutman. Graphs with smallest sum of squares of vertex degrees. *Kragujevac Journal of Mathematics*, 25:51–54, 2003 (cited on page 70).

[Gut05]   Ivan Gutman. What chemists could not see without mathematics – dependence of total π-electron energy on molecular structure. *Kragujevac Journal of Science*, 27:57–66, 2005 (cited on page 155).

[Gut11]   Ivan Gutman. Multiplicative Zagreb indices of trees. *Bulletin of the International Mathematical Virtual Institute*, 1(1):13–19, 2011 (cited on page 80).

[Gut13]   Ivan Gutman. Degree-based topological indices. *Croatica Chemica Acta*, 86(4):351–361, December 2013 (cited on page 30).

[Gut14a]   Ivan Gutman. An exceptional property of first Zagreb index. *MATCH: Communications in Mathematical and in Computer Chemistry*, 72(3):733–740, 2014 (cited on page 70).

[Gut14b]   Ivan Gutman. On the origin of two degree-based topological indices. *Bulletin de l'Académie serbe des sciences et des arts, Classe des Sciences mathématiques et naturelles, Sciences mathématiques*, CXLVI(39):39–52, 2014 (cited on page 29).

[Gut78]   Ivan Gutman. The energy of a graph. Berichte der Mathematisch-Statistischen Sektion 103, Forschungszentrum Graz, September 1978 (cited on page 154).

[Hat10]   Hamed Hatami. Graph norms and Sidorenko's conjecture. *Israel Journal of Mathematics*, 175(1):125–150, January 2010 (cited on page 163).

[Hay20]   Tsuruichi Hayashi. On some inequalities. *Rendiconti del Circolo Matematico di Palermo*, 44:336–340, 1920 (cited on page 15).

[HB78]   Robert L. Hemminger and Lowell W. Beineke. Line graphs and line digraphs. In *Selected Topics in Graph Theory*, pages 271–305. Academic Press, 1978 (cited on page 31).

[Her55]   Charles Hermite. Remarque sur un théorème de M. Cauchy. *Comptes rendus hebdomadaires des séances de l'Académie des sciences*, 41:181–183, 1855 (cited on page 39).

[HJ13]   Roger A. Horn and Charles R. Johnson. *Matrix Analysis*. Cambridge University Press, 2nd edition, 2013 (cited on pages 39, 136, 138, 139, 150).

[HJ91]   Roger A. Horn and Charles R. Johnson. *Topics in Matrix Analysis*. Cambridge University Press, 1991 (cited on pages 112, 160).

[HKM⁺11]   Raymond Hemmecke, Sven Kosub, Ernst W. Mayr, Hanjo Täubig and Jeremias Weihmann. Inequalities for the Number of Walks in Trees and General Graphs and a Generalization of a Theorem of Erdős and Simonovits. Technical report TUM-I1109, Department of Computer Science, Technische Universität München, April 2011 (cited on pages xiii, 48).

[HKM⁺12]   Raymond Hemmecke, Sven Kosub, Ernst W. Mayr, Hanjo Täubig and Jeremias Weihmann. Inequalities for the number of walks in graphs. In *Proceedings of the 9th Meeting on Analytic Algorithmics and Combinatorics (ANALCO'12)*, pages 26–39. SIAM, January 2012 (cited on pages xiii, 48, 92).

[HKMT06]   Klaus Holzapfel, Sven Kosub, Moritz G. Maaß and Hanjo Täubig. The complexity of detecting fixed-density clusters. *Discrete Applied Mathematics*, 154(11):1547–1562, July 2006 (cited on page 22).

[HLP59]   Godfrey H. Hardy, John E. Littlewood and George Pólya. *Inequalities*. Cambridge University Press, 2nd edition, 1959 (cited on pages 14, 18).

[HMV86]   Seppo Hyyrö, Jorma Kaarlo Merikoski and Ari Virtanen. Improving certain simple eigenvalue bounds. *Mathematical Proceedings of the Cambridge Philosophical Society*, 99(3):507–518, May 1986 (cited on pages 63, 64, 139, 141, 152).

[HN60]   Frank Harary and Robert Z. Norman. Some properties of line digraphs. *Rendiconti del Circolo Matematico di Palermo*, 9(2):161–168, 1960 (cited on page 31).

[HO07]   Akihito Hora and Nobuaki Obata. *Quantum Probability and Spectral Analysis of Graphs, Theoretical and Mathematical Physics*. Springer, 2007 (cited on page 34).

[Hof67]   Alan J. Hoffman. Three observations on nonnegative matrices. *Journal of Research of the National Bureau of Standards*, 71B(1):39–41, March 1967 (cited on pages 71, 72).

[Hof70]   Alan J. Hoffman. On eigenvalues and colorings of graphs. In Bernard Harris, editor, *Graph Theory and Its Applications*, pages 79–91. Academic Press, 1970 (cited on page 27).

[Hof88]   Michael Hofmeister. Spectral radius and degree sequence. *Mathematische Nachrichten*, 139(1):37–44, 1988 (cited on pages 151, 156).

[Hol05]   Boris Hollas. On the variance of topological indices that depend on the degree of a vertex. *MATCH: Communications in Mathematical and in Computer Chemistry*, 54(2):341–350, 2005 (cited on page 25).

[Hol06]   Boris Hollas. An analysis of the redundancy of graph invariants used in chemoinformatics. *Discrete Applied Mathematics*, 154(17):2484–2498, November 2006 (cited on page 25).

[Höl89]   Otto Hölder. Ueber einen Mittelwerthssatz. *Nachrichten von der Königl. Gesellschaft der Wissenschaften und der Georg-Augusts-Universität zu Göttingen*, 1889(2):38–47, January 1889 (cited on page 14).

[Hon93]   Yuan Hong. Bounds of eigenvalues of graphs. *Discrete Mathematics*, 123(1-3):65–74, December 1993 (cited on page 151).

[HOS92]   Marie-Claude Heydemann, Jaroslav Opatrny and Dominique Sotteau. Broadcasting and spanning trees in de Bruijn and Kautz networks. *Discrete Applied Mathematics*, 37-38:297–317, July 1992 (cited on page 32).

[HR14]   Asma Hamzeh and Tamás Réti. An analogue of Zagreb index inequality obtained from graph irregularity measures. *MATCH: Communications in Mathematical and in Computer Chemistry*, 72(3):669–683, 2014 (cited on page 25).

[HS79]   Frank Harary and Allen J. Schwenk. The spectral approach to determining the number of walks in a graph. *Pacific Journal of Mathematics*, 80(2):443–449, 1979 (cited on page 38).

[HTW07]   Yaoping Hou, Zhen Tang and Chingwah Woo. On the spectral radius, k-degree and the upper bound of energy in a graph. *MATCH: Communications in Mathematical and in Computer Chemistry*, 57(2):341–350, 2007 (cited on pages 151, 156).

[Hu09]   Shengbiao Hu. A sharp lower bound of the spectral radius of simple graphs. *Applicable Analysis and Discrete Mathematics*, 3(2):379–385, October 2009 (cited on page 151).

[HV07]   Pierre Hansen and Damir Vukičević. Comparing the Zagreb indices. *Croatica Chemica Acta*, 80(2):165–168, June 2007 (cited on page 85).

[HZ05]   Yuan Hong and Xiao-Dong Zhang. Sharp upper and lower bounds for largest eigenvalue of the Laplacian matrices of trees. *Discrete Mathematics*, 296(2-3):187–197, July 2005 (cited on pages 151, 156).

[Ili12]   Aleksandar Ilić. Note on the harmonic index of a graph, April 2012. arXiv: 1204.3313v1 [math.CO] (cited on page 83).

[IS09]   Aleksandar Ilić and Dragan Stevanović. On comparing Zagreb indices. *MATCH: Communications in Mathematical and in Computer Chemistry*, 62(3):681–687, 2009 (cited on pages 60, 64, 65, 100).

[Juk11]   Stasys Jukna. *Extremal combinatorics – with applications in computer science*. In Texts in Theoretical Computer Science. An EATCS Series. Springer, 2011. Part 15. Eigenvalues and Graph Expansion (cited on page 27).

[Kar93]   Alan F. Karr. *Probability, Springer Texts in Statistics*. Springer, 2nd edition, 1993 (cited on page 68).

[Kas63]   Pieter W. Kasteleyn. A soluble self-avoiding walk problem. *Physica*, 29(12):1329–1337, December 1963 (cited on page 31).

[Kas67]   Pieter W. Kasteleyn. Graph theory and crystal physics. In Frank Harary, editor, *Graph Theory and Theoretical Physics*, pages 43–110. Academic Press, 1967 (cited on pages 32, 34).

[Kau68]   William H. Kautz. Bounds on directed (d, k) graphs. In *Theory of cellular logic networks and machines. AFCRL-68-0668, SRI Project 7258 Final Report*, pages 20–28. December 1968 (cited on page 32).

[Kau71]   William H. Kautz. Design of optimal interconnection networks for multiprocessors. In *Structure et conception des ordinateurs / Architecture and design of digital computers. École d'été de l'OTAN / NATO Advanced Summer Institute, 1969*, pages 249–272. Dunod, Paris, 1971 (cited on page 32).

[Khi32]   Aleksandr Khintchine. Über eine Ungleichung. *Математический сборник (Matematiceskij Sbornik)*, 39:35–39, 1932 (cited on pages 123, 124).

[Khi33]   Aleksandr Khintchine. Über ein metrisches Problem der additiven Zahlentheorie. *Математический сборник (Matematiceskij Sbornik)*, 40:180–189, 1933 (cited on pages 34, 123).

[Kin61a]   John F. C. Kingman. A mathematical problem in population genetics. *Proceedings of the Cambridge Philosophical Society*, 57(3):574–582, July 1961 (cited on page 28).

[Kin61b]   John F. C. Kingman. On an inequality in partial averages. *The Quarterly Journal of Mathematics (Oxford Second Series)*, 12:78–80, 1961 (cited on pages 28, 110).

[Kin67]   John F. C. Kingman. An inequality involving Radon-Nikodym derivatives. *Proceedings of the Cambridge Philosophical Society*, 63(1):195–198, January 1967 (cited on page 110).

[KK03]   M. Frans Kaashoek and David R. Karger. Koorde: a simple degree-optimal distributed hash table. In *Proceedings of the 2nd International Workshop on Peer-to-Peer Systems (IPTPS'03)*, volume 2735 of *Lecture Notes in Computer Science*, pages 98–107, February 2003 (cited on page 32).

[KLL16]   Jeong Han Kim, Choongbum Lee and Joonkyung Lee. Two approaches to Sidorenko's conjecture. *Transactions of the American Mathematical Society*, 368(7):5057–5074, July 2016 (cited on page 164).

[KM01]   Jack H. Koolen and Vincent Moulton. Maximal energy graphs. *Advances in Applied Mathematics*, 26(1):47–52, January 2001 (cited on pages 155, 156).

[KM03]   Jack H. Koolen and Vincent Moulton. Maximal energy bipartite graphs. *Graphs and Combinatorics*, 19(1):131–135, March 2003 (cited on page 155).

[KM84]   Hilkka Kankaanpää and Jorma Kaarlo Merikoski. On a conjecture of D. London. *Linear and Multilinear Algebra*, 15(2):171–173, May 1984 (cited on page 74).

[KMG00]   Jack H. Koolen, Vincent Moulton and Ivan Gutman. Improving the McClelland inequality for total π-electron energy. *Chemical Physics Letters*, 320(3-4):213–216, April 2000 (cited on pages 155, 156).

[Kna11]   Ulrich Knauer. *Algebraic Graph Theory. Morphisms, Monoids and Matrices*, volume 41 of *Studies in Mathematics*. de Gruyter, 2011 (cited on pages 38, 135).

[Knu67]   Donald E. Knuth. Oriented subtrees of an arc digraph. *Journal of Combinatorial Theory*, 3(4):309–314, December 1967 (cited on page 31).

[Kol04]   Lilia Yu. Kolotilina. Bounds and inequalities for the Perron root of a nonnegative matrix. *Journal of Mathematical Sciences*, 121(4):2481–2507, June 2004 (cited on page 145).

[Kol05]   Lilia Yu. Kolotilina. Bounds and inequalities for the Perron root of a nonnegative matrix. II. Circuit bounds and inequalities. *Journal of Mathematical Sciences*, 127(3):1988–2005, May 2005 (cited on page 145).

[Kol06]   Lilia Yu. Kolotilina. Bounds for the Perron root, singularity/nonsingularity conditions, and eigenvalue inclusion sets. *Numerical Algorithms*, 42(3-4):247–280, July 2006 (cited on page 145).

[Kol93]   Lilia Yu. Kolotilina. Lower bounds for the Perron root of a nonnegative matrix. *Linear Algebra and its Applications*, 180:133–151, February 1993 (cited on page 145).

[Kos05]   Sven Kosub. Local density. In Ulrik Brandes and Thomas Erlebach, editors, *Network Analysis. Methodological Foundations*. Volume 3418, Lecture Notes in Computer Science, pages 112–142. Springer, 2005 (cited on pages 22, 57, 66).

[KP93]   Guy Kortsarz and David Peleg. On choosing a dense subgraph. In *Proceedings of the 34th Annual Symposium on Foundations of Computer Science (FOCS'93)*, pages 692–701. IEEE CS, November 1993 (cited on page 22).

[KPRT04]   Douglas J. Klein, José Luis Palacios, Milan Randić and Nenad Tri-
           najstić. Random walks and chemical graph theory. *Journal of Chem-
           ical Information and Computer Sciences*, 44(5):1521–1525, Septem-
           ber 2004 (cited on page 31).

[Kra43]   József Krausz. Egy Whitney-féle gráf-tétel új bizonyítása (Démons-
           tration nouvelle d'une théorème de Whitney sur les réseaux). *Matem-
           atikai és Fizikai Lapok*, 50:75–85, 1943 (cited on page 31).

[KS96]   János Komlós and Miklós Simonovits. Szemerédi's regularity lemma
           and its applications in graph theory. In D. Miklós, V. T. Sós and
           T. Szőnyi, editors, *Combinatorics, Paul Erdős is Eighty*. Volume 2,
           pages 295–352. János Bolyai Mathematical Society, 1996 (cited on
           page 22).

[KSV13]   Peter Keevash, Benny Sudakov and Jacques Verstraëte. On a conjec-
           ture of Erdős and Simonovits: even cycles. *Combinatorica*, 33(6):699–
           732, December 2013 (cited on page 24).

[Kwa96]   Jaroslaw Kwapisz. On the spectral radius of a directed graph.
           Journal of Graph Theory, 23(4):405–411, December 1996 (cited on
           pages 8, 31–33, 148, 149).

[Law86]   Gregory F. Lawler. Expected hitting times for a random walk on a
           connected graph. *Discrete Mathematics*, 61(1):85–92, August 1986
           (cited on page 26).

[Lev11]   Lionel Levine. Sandpile groups and spanning trees of directed line
           graphs. *Journal of Combinatorial Theory, Series A*, 118(2):350–364,
           February 2011 (cited on page 33).

[Li09]   Rao Li. The spectral moments and energy of graphs. *Applied Math-
           ematical Sciences*, 3(56):2765–2773, 2009 (cited on page 158).

[Liu96]   Shu-Lin Liu. Bounds for the greatest characteristic root of a non-
           negative matrix. *Linear Algebra and its Applications*, 239:151–160,
           May 1996 (cited on page 145).

[LL78]   Raphael Loewy and David London. A note on an inverse problem
           for nonnegative matrices. *Linear and Multilinear Algebra*, 6(1):83–
           90, January 1978 (cited on page 143).

[LLT07]   Huiqing Liu, Mei Lu and Feng Tian. Some upper bounds for the
           energy of graphs. *Journal of Mathematical Chemistry*, 41(1):45–57,
           January 2007 (cited on page 156).

[LMNT02]   István Lukovits, Ante Miličević, Sonja Nikolić and Nenad Trinajstić.
           On walk counts and complexity of general graphs. *Internet Electronic
           Journal of Molecular Design*, 1(8):388–400, August 2002 (cited on
           page 30).

[LMSM83]   Jeffrey C. Lagarias, James E. Mazo, Lawrence A. Shepp and Bren-
           dan D. McKay. An inequality for walks in a graph. *SIAM Review*,
           25(3):403, July 1983 (cited on page 47).

[LMSM84]   Jeffrey C. Lagarias, James E. Mazo, Lawrence A. Shepp and Bren-
           dan D. McKay. An inequality for walks in a graph. *SIAM Review*,
           26(4):580–582, October 1984 (cited on pages 47–54, 74, 75, 85–87).

[Lon66a]   David London. Inequalities in quadratic forms. *Duke Mathematical Journal*, 33(3):511–522, September 1966 (cited on pages 22, 58).

[Lon66b]   David London. Two inequalities in nonnegative symmetric matrices. *Pacific Journal of Mathematics*, 16(3):515–536, 1966 (cited on pages 71, 72, 74, 75).

[Lot83]   M. Lothaire. *Combinatorics on Words*, volume 17 of *Encyclopedia of Mathematics and its Applications*. Addison-Wesley, 1983 (cited on page 21).

[Lov09]   László Lovász. Very large graphs. *Current Developments in Mathematics*, 2008:67–128, 2009 (cited on page 164).

[Lov12]   László Lovász. *Large Networks and Graph Limits*, volume 60 of *Colloquium Publications*. American Mathematical Society, 2012 (cited on page 164).

[Lov79]   László Lovász. *Combinatorial Problems and Exercises*. Akadémiai Kiadó (Budapest), North-Holland (New York / Oxford), 1979 (cited on page 30).

[Lov96]   László Lovász. Random walks on graphs: A survey. In D. Miklós, V. T. Sós and T. Szőnyi, editors, *Combinatorics, Paul Erdős is Eighty*. Volume 2, pages 353–397. János Bolyai Mathematical Society, 1996 (cited on page 26).

[LP01]   Jiong-Sheng Li and Yong-Liang Pan. De Caen's inequality and bounds on the largest Laplacian eigenvalue of a graph. *Linear Algebra and its Applications*, 328:153–160, May 2001 (cited on page 70).

[LP82]   Harry R. Lewis and Christos H. Papadimitriou. Symmetric space-bounded computation. *Theoretical Computer Science*, 19(2):161–187, August 1982 (cited on page 21).

[LQW+15]   Xueliang Li, Zhongmei Qin, Meiqin Wei, Ivan Gutman and Matthias Dehmer. Novel inequalities for generalized graph entropies – graph energies and topological indices. *Applied Mathematics and Computation*, 259:470–479, May 2015 (cited on page 34).

[LS11]   J. L. Xiang Li and Balázs Szegedy. On the logarithmic calculus and Sidorenko's conjecture, July 2011. arXiv: 1107.1153v1 [math.CO] (cited on page 164).

[LSG12]   Xueliang Li, Yongtang Shi and Ivan Gutman. *Graph Energy*. Springer, 2012 (cited on pages 155, 157).

[LT03]   István Lukovits and Nenad Trinajstić. Atomic walk counts of negative order. *Journal of Chemical Information and Computer Sciences*, 43(4):1110–1114, July 2003 (cited on page 31).

[LTV13]   Dajie Liu, Stojan Trajanovski and Piet Van Mieghem. Random line graphs and a linear law for assortativity. *Physical Review E*, 87(1):012816, January 2013 (cited on page 34).

[Lux72]   Wilhelmus A. J. Luxemburg. On an inequality of A. Khintchine for zero-one matrices. *Journal of Combinatorial Theory, Series A*, 12(2):289–296, March 1972 (cited on page 124).

[Lya01]   Aleksandr M. Lyapunov. Nouvelle forme du théorème sur la limite de probabilité. *Mémoires de l'Académie Impériale des sciences de St.-Pétersbourg, VIIIe Série*, 12(5):1–24, 1901 (cited on page 14).

[Man68]   S. P. H. Mandel. Fundamental theorem of natural selection. *Nature*, 220(5173):1251–1252, December 1968 (cited on page 28).

[Man80]   S. P. H. Mandel. The triangle inequality in balanced genetical polymorphisms. *Genetics*, 96(2):557–559, October 1980 (cited on page 28).

[Mat90]   František Matúš. Inequalities concerning the demi-degrees and numbers of paths. Research Report 1652, ÚTIA ČSAV, Prague, Czech Republic, April 1990 (cited on pages 124, 125).

[McC71]   Bernard J. McClelland. Properties of the latent roots of a matrix: the estimation of π-electron energies. *The Journal of Chemical Physics*, 54(2):640–643, January 1971 (cited on pages 154, 155, 158).

[MH58]   S. P. H. Mandel and I. M. Hughes. Change in mean viability at a multiallelic locus in a population under random mating. *Nature*, 182(4627):63–64, July 1958 (cited on pages 28, 110).

[Min88]   Henryk Minc. *Nonnegative Matrices*. John Wiley & Sons, 1988 (cited on pages 27, 123, 142, 144, 145).

[Mit64]   Dragoslav S. Mitrinović. *Elementary Inequalities*. P. Noordhoff Ltd., Groningen, The Netherlands, 1964 (cited on page 18).

[Mit70]   Dragoslav S. Mitrinović. *Analytic Inequalities*, volume 165 of *Die Grundlehren der mathematischen Wissenschaften*. Springer, 1970 (cited on page 18).

[MM64]   Marvin Marcus and Henryk Minc. *A Survey of Matrix Theory and Matrix Inequalities*. Allyn and Bacon, Inc., Boston, 1964 (cited on pages 18, 145).

[MN62]   Marvin Marcus and Morris Newman. The sum of the elements of the powers of a matrix. *Pacific Journal of Mathematics*, 12(2):627–635, 1962 (cited on pages 49, 53, 54, 74, 109, 141).

[Mor65]   H. L. Morgan. The generation of a unique machine description for chemical structures—A technique developed at chemical abstracts service. *Journal of Chemical Documentation*, 5(2):107–113, May 1965 (cited on page 30).

[MP93]   Bojan Mohar and Svatopluk Poljak. Eigenvalues in combinatorial optimization. In *Combinatorial and Graph-Theoretical Problems in Linear Algebra*. Volume 50, The IMA Volumes in Mathematics and its Applications, pages 107–151. Springer, 1993 (cited on page 27).

[MPF91]   Dragoslav S. Mitrinović, Josip E. Pečarić and Arlington M. Fink. *Inequalities Involving Functions and Their Integrals and Derivatives*, volume 53 of *Mathematics and Its Applications (East European Series)*. Kluwer Academic Publishers, 1991 (cited on page 160).

[MPF93]   Dragoslav S. Mitrinović, Josip E. Pečarić and Arlington M. Fink. *Classical and New Inequalities in Analysis*, volume 61 of *Mathematics and Its Applications (East European Series)*. Kluwer Academic Publishers, 1993 (cited on page 18).

[MRS12]   Marko Milošević, Tamás Réti and Dragan Stevanović. On the constant difference of Zagreb indices. *MATCH: Communications in Mathematical and in Computer Chemistry*, 68(1):157–168, 2012 (cited on page 30).

[MS59]   H. P. Mulholland and Cedric A. B. Smith. An inequality arising in genetical theory. *The American Mathematical Monthly*, 66(8):673–683, October 1959 (cited on pages 22, 24, 28, 58, 60, 61, 68, 69, 110, 111, 114, 160, 163).

[MS60]   H. P. Mulholland and Cedric A. B. Smith. Corrections: an inequality arising in genetical theory. *The American Mathematical Monthly*, 67(2):161, February 1960 (cited on page 60).

[MT92]   František Matúš and Antonín Tuzar. Short proofs of Khintchine-type inequalities for zero-one matrices. *Journal of Combinatorial Theory, Series A*, 59(1):155–159, January 1992 (cited on page 124).

[MV74]   Dragoslav S. Mitrinović and Petar M. Vasić. History, variations and generalisations of the čebyšev inequality and the question of some priorities. *Univ. Beograd, Publ. Elektrotehn. Fak., Ser. Mat. Fiz.*, 461–497(461):1–30, 1974 (cited on page 15).

[MV91]   Jorma Kaarlo Merikoski and Ari Virtanen. On the London-Hoffman inequality for the sum of elements of powers of nonnegative matrices. *Linear and Multilinear Algebra*, 30(4):257–259, November 1991 (cited on page 129).

[MV95]   Jorma Kaarlo Merikoski and Ari Virtanen. Research problem: an inequality for the sum of elements of powers of nonnegative matrices. *Linear and Multilinear Algebra*, 39(3):307–308, August 1995 (cited on pages 127–129).

[MW89]   Bojan Mohar and Wolfgang Woess. A survey on spectra of infinite graphs. *Bulletin of the London Mathematical Society*, 21(3):209–234, May 1989 (cited on page 38).

[New02]   Mark E. J. Newman. Assortative mixing in networks. *Physical Review Letters*, 89(20):208701, October 2002 (cited on page 34).

[New03]   Mark E. J. Newman. Mixing patterns in networks. *Physical Review E*, 67(2):026126, February 2003 (cited on page 34).

[Nik02]   Vladimir Nikiforov. Some inequalities for the largest eigenvalue of a graph. *Combinatorics, Probability and Computing*, 11(2):179–189, March 2002 (cited on page 152).

[Nik06a]   Vladimir Nikiforov. Eigenvalues and degree deviation in graphs. *Linear Algebra and its Applications*, 414(1):347–360, April 2006 (cited on page 25).

[Nik06b]   Vladimir Nikiforov. Walks and the spectral radius of graphs. *Linear Algebra and its Applications*, 418(1):257–268, October 2006 (cited on pages 89, 151–153).

[Nik07a]   Vladimir Nikiforov. Graphs and matrices with maximal energy. *Journal of Mathematical Analysis and Applications*, 327(1):735–738, March 2007 (cited on page 154).

[Nik07b]   Vladimir Nikiforov. Revisiting Schur's bound on the largest singular value. Technical report math/0702722v2, arXiv, March 2007 (cited on page 138).

[Nik07c]   Vladimir Nikiforov. The energy of graphs and matrices. *Journal of Mathematical Analysis and Applications*, 326(2):1472–1475, February 2007 (cited on page 154).

[Nik07d]   Vladimir Nikiforov. The sum of the squares of degrees: sharp asymptotics. *Discrete Mathematics*, 307(24):3187–3193, November 2007 (cited on page 70).

[Nik09]   Vladimir Nikiforov. More spectral bounds on the clique and independence numbers. *Journal of Combinatorial Theory, Series B*, 99(6):819–826, November 2009 (cited on page 152).

[Nik11]   Vladimir Nikiforov. Some new results in extremal graph theory. In *Surveys in Combinatorics*. Volume 392, London Mathematical Society Lecture Note Series, pages 141–181. Cambridge University Press, 2011 (cited on pages 24, 152).

[Nik15]   Vladimir Nikiforov. Beyond graph energy: norms of graphs and matrices, October 2015. arXiv: 1510 . 02850v1 [math.CO] (cited on page 154).

[Nin14]   Bo Ning. On some papers of Nikiforov, September 2014. arXiv: 1409 . 5882v1 [math.CO] (cited on page 152).

[NKMT03]   Sonja Nikolić, Goran Kovačević, Ante Miličević and Nenad Trinajstić. The Zagreb indices 30 years after. *Croatica Chemica Acta*, 76(2):113–124, June 2003 (cited on pages 29, 30).

[Nos70]   Eva Nosal. *Eigenvalues of graphs*. Master's thesis, University of Calgary, 1970 (cited on page 153).

[NW03a]   Moni Naor and Udi Wieder. A simple fault tolerant distributed hash table. In *Proceedings of the 2ⁿᵈ International Workshop on Peer-to-Peer Systems (IPTPS'03)*, volume 2735 of *Lecture Notes in Computer Science*, pages 88–97, February 2003 (cited on page 32).

[NW03b]   Moni Naor and Udi Wieder. Novel architectures for P2P applications: the continuous-discrete approach. In *Proceedings of the 15ᵗʰ ACM Symposium on Parallelism in Algorithms and Architectures (SPAA'03)*, pages 50–59. ACM, June 2003 (cited on page 32).

[NW07]   Moni Naor and Udi Wieder. Novel architectures for P2P applications: the continuous-discrete approach. *ACM Transactions on Algorithms*, 3(3):Article 34, August 2007 (cited on page 32).

[Oba04]   Nobuaki Obata. Quantum probabilistic approach to spectral analysis of star graphs. *Interdisciplinary Information Sciences*, 10(1):41–52, 2004 (cited on page 34).

[Pal90]   José Luis Palacios. Bounds on expected hitting times for a random walk on a connected graph. *Linear Algebra and its Applications*, 141:241–252, November 1990 (cited on page 26).

[Par98]   Beresford N. Parlett. *The Symmetric Eigenvalue Problem*, volume 20 of *Classics in Applied Mathematics*. SIAM, 1998 (cited on page 135).

[Pat12]   Thomas H. Pate. Extending the Hölder type inequality of Blakley and Roy to non-symmetric non-square matrices. *Transactions of the American Mathematical Society*, 364(8):4267–4281, August 2012 (cited on pages 109, 111, 113, 163, 164).

[PCAC14]   Diego H. Peluffo-Ordóñez, Cristian Castro-Hoyos, Carlos D. Acosta-Medina and Germán Castellanos-Domínguez. Quadratic problem formulation with linear constraints for normalized cut clustering. In *Progress in Pattern Recognition, Image Analysis, Computer Vision, and Applications / Proceedings of the 19th Iberoamerican Congress on Pattern Recognition (CIARP'14)*, volume 8827 of *Lecture Notes in Computer Science*, pages 408–415, November 2014 (cited on page 111).

[Pea80]   Lynn Hauser Pearce. Random walks on trees. *Discrete Mathematics*, 30(3):269–276, 1980 (cited on page 26).

[Pen65]   Robert H. Penny. A connectivity code for use in describing chemical structures. *Journal of Chemical Documentation*, 5(2):113–117, May 1965 (cited on page 30).

[Per01]   Dominique Perrin. Enumerative combinatorics on words. In *Algebraic Combinatorics and Computer Science*, pages 391–427. Springer, 2001 (cited on page 21).

[Per07]   Oskar Perron. Zur Theorie der Matrices. *Mathematische Annalen*, 64(2):248–263, 1907 (cited on page 142).

[Pev89]   Pavel A. Pevzner. *l*-tuple DNA sequencing: computer analysis. *Journal of Biomolecular Structure and Dynamics*, 7(1):63–73, 1989 (cited on page 32).

[Pla47]   John R. Platt. Influence of neighbor bonds on additive bond properties in paraffins. *The Journal of Chemical Physics*, 15(6):419–420, June 1947 (cited on page 29).

[Pla52]   John R. Platt. Prediction of isomeric differences in paraffin properties. *The Journal of Physical Chemistry*, 56(3):328–336, March 1952 (cited on page 29).

[PPS99]   Uri N. Peled, Rossella Petreschi and Andrea Sterbini. (n, e)-graphs with maximum sum of squares of degrees. *Journal of Graph Theory*, 31(4):283–295, August 1999 (cited on page 70).

[PT01]   Pavel A. Pevzner and Haixu Tang. Fragment assembly with double-barreled data. *Bioinformatics*, 17(Suppl.1):S225–S233, June 2001 (cited on page 32).

[PTW01]   Pavel A. Pevzner, Haixu Tang and Michael S. Waterman. An Eulerian path approach to DNA fragment assembly. *Proceedings of the National Academy of Sciences of the United States of America*, 98(17):9748–9753, August 2001 (cited on page 32).

[Raz82]   Marko Razinger. Extended connectivity in chemical graphs. *Theoretica Chimica Acta*, 61(6):581–586, September 1982 (cited on page 30).

[Raz86]   Marko Razinger. Discrimination and ordering of chemical structures by the number of walks. *Theoretica Chimica Acta*, 70(5):365–378, November 1986 (cited on page 30).

[RG12]   Tamás Réti and Ivan Gutman. Relations between ordinary and multiplicative Zagreb indices. *Bulletin of the International Mathematical Virtual Institute*, 2(2):133–140, 2012 (cited on page 80).

[RHJ88]   Milan Randić, Peter J. Hansen and Peter C. Jurs. Search for useful graph theoretical invariants of molecular structure. *Journal of Chemical Information and Computer Sciences*, 28(2):60–68, May 1988 (cited on page 30).

[Rog88]   Leonard J. Rogers. An extension of a certain theorem in inequalities. *Messenger of Mathematics*, XVII(10):145–150, February 1888 (cited on page 14).

[RR00]   Christoph Rücker and Gerta Rücker. Walk counts, labyrinthicity, and complexity of acyclic and cyclic graphs and molecules. *Journal of Chemical Information and Computer Sciences*, 40(1):99–106, January 2000 (cited on page 30).

[RR01]   Christoph Rücker and Gerta Rücker. Substructure, subgraph, and walk counts as measures of the complexity of graphs and molecules. *Journal of Chemical Information and Computer Sciences*, 41(6):1457–1462, November 2001 (cited on page 30).

[RR03]   Christoph Rücker and Gerta Rücker. Walking backward: walk counts of negative order. *Journal of Chemical Information and Computer Sciences*, 43(4):1115–1120, July 2003 (cited on page 31).

[RR93]   Christoph Rücker and Gerta Rücker. Counts of all walks as atomic and molecular descriptors. *Journal of Chemical Information and Computer Sciences*, 33(5):683–695, September 1993 (cited on page 30).

[Sab68]   Gert Sabidussi. Existenz und Struktur selbstadjungierter Graphen. In *Beiträge zur Graphentheorie*, pages 121–125. B. G. Teubner, Leipzig, 1968 (cited on page 31).

[SCE92]   László A. Székely, L. H. Clark and Roger C. Entringer. An inequality for degree sequences. *Discrete Mathematics*, 103(3):293–300, May 1992 (cited on page 70).

[Sch11]   Issai Schur. Bemerkungen zur Theorie der beschränkten Bilinearformen mit unendlich vielen Veränderlichen. *Journal für die reine und angewandte Mathematik*, 140:1–28, 1911 (cited on page 137).

[Sch64]   Binyamin Schwarz. Rearrangements of square matrices with non-negative elements. *Duke Mathematical Journal*, 31(1):45–62, March 1964 (cited on page 147).

[Sch88]   Hermann A. Schwarz. Ueber ein die Flächen kleinsten Flächeninhalts betreffendes Problem der Variationsrechnung. Festschrift zum Jubelgeburtstage des Herrn Karl Weierstrass. *Acta Societatis Scientiarum Fennicae*, XV:315–362, 1888 (cited on page 13).

[SD00]   József Sándor and Lokenath Debnath. On certain inequalities involving the constant e and their applications. *Journal of Mathematical Analysis and Applications*, 249(2):569–582, September 2000 (cited on page 132).

[SD88]   V. A. Skorobogatov and A. A. Dobrynin. Metric analysis of graphs. *MATCH: Communications in Mathematical and in Computer Chemistry*, 23:105–151, 1988 (cited on page 30).

[Sen73]   Eugene Seneta. *Non-negative Matrices – An Introduction to Theory and Applications*. George Allen & Unwin Ltd, London, 1973 (cited on pages 27, 123).

[Sid85a]   Alexander F. Sidorenko. Proof of London's conjecture on sums of elements of positive matrices. *Mathematical Notes of the Academy of Sciences of the USSR*, 38(3):716–717, September 1985 (cited on page 72).

[Sid85b]   Alexander F. Sidorenko. Доказательство предположения Лондона о суммах элементов неотрицательных матриц. *Математические заметки (Matematicheskie Zametki)*, 38(3):376–377, 1985 (cited on page 72).

[Sid91]   Alexander F. Sidorenko. Неравенства для функционалов, порождаемых двудольными графами. *Дискретная математика (Diskretnaya Matematika)*, 3(3):50–65, 1991 (cited on page 162).

[Sid92]   Alexander F. Sidorenko. Inequalities for functionals generated by bipartite graphs. *Discrete Mathematics and Applications*, 2(5):489–504, January 1992 (cited on pages 162, 163).

[Sid93]   Alexander F. Sidorenko. A correlation inequality for bipartite graphs. *Graphs and Combinatorics*, 9(2-4):201–204, June 1993 (cited on pages 111, 163).

[SM59]   P. A. G. Scheuer and S. P. H. Mandel. An inequality in population genetics. *Heredity*, 13(4):519–524, November 1959 (cited on pages 22, 28, 58, 60).

[Sne08]   Jan Snellman. Digraphs with a fixed number of edges and vertices, having a maximal number of walks of length 2, April 2008. arXiv: 0804.4655v1 [math.CO] (cited on pages 31, 32).

[SP89]   Maheswara R. Samatham and Dhiraj K. Pradhan. The de Bruijn multiprocessor network: a versatile parallel processing and sorting network for VLSI. *IEEE Transactions on Computers*, 38(4):567–581, April 1989 (cited on page 32).

[SP91]   Maheswara R. Samatham and Dhiraj K. Pradhan. Correction to "The de Bruijn multiprocessor network: a versatile parallel processing and sorting network for VLSI". *IEEE Transactions on Computers*, 40(1):122–123, January 1991 (cited on page 32).

[Spi12]   Daniel Spielman. Spectral graph theory. In Uwe Naumann and Olaf Schenk, editors, *Combinatorial Scientific Computing*, part 16, pages 495–524. Chapman & Hall / CRC Press, 2012 (cited on pages 27, 38).

[Spi64a]   Leonard Spialter. The atom connectivity matrix (ACM) and its characteristic polynomial (ACMCP). *Journal of Chemical Documentation*, 4(4):261–269, October 1964 (cited on page 29).

[Spi64b]   Leonard Spialter. The atom connectivity matrix characteristic polynomial (ACMCP) and its physico-geometeric (topological) significance. *Journal of Chemical Documentation*, 4(4):269–274, October 1964 (cited on page 29).

[SR94]   Kumar N. Sivarajan and Rajiv Ramaswami. Lightwave networks based on de Bruijn graphs. *IEEE/ACM Transactions on Networking*, 2(1):70–79, February 1994 (cited on page 32).

[SS78]   Arto Salomaa and Matti Soittola. *Automata-Theoretic Aspects of Formal Power Series, Texts and Monographs in Computer Science*. Springer, 1978 (cited on page 21).

[ST48]   Herbert Seifert and William Threlfall. Topologie. In *Naturforschung und Medizin in Deutschland 1939–1946 (Für Deutschland bestimmte Ausgabe der Fiat Review of German Science)*. Volume 2, pages 239–252. Dieterich'sche Verlagsbuchhandlung, Wiesbaden, 1948 (cited on page 37).

[Sta13]   Richard P. Stanley. *Algebraic Combinatorics. Walks, Trees, Tableaux, and More, Undergraduate Texts in Mathematics*. Springer, 2013 (cited on page 8).

[Sta87]   Richard P. Stanley. A bound on the spectral radius of graphs with e edges. *Linear Algebra and its Applications*, 87:267–269, March 1987 (cited on page 151).

[Ste04]   J. Michael Steele. *The Cauchy-Schwarz Master Class. An Introduction to the Art of Mathematical Inequalities, MAA Problem Books Series*. Cambridge University Press, 2004 (cited on pages 18, 130).

[Ste11]   Dragan Stevanović. Applications of graph spectra in quantum physics. In *Selected Topics on Applications of Graph Spectra*. Volume 14(22), Zbornik radova, pages 85–111. Mathematical Institute SANU, Belgrade, January 2011 (cited on page 34).

[Ste15]   Dragan Stevanović. Walk counts and the spectral radius of graphs. *Bulletin de l'Académie serbe des sciences et des arts, Classe des Sciences mathématiques et naturelles, Sciences mathématiques*, CXLVIII(40):33–57, 2015 (cited on page 43).

[Ste67]   George Stell. Generating functionals and graphs. In Frank Harary, editor, *Graph Theory and Theoretical Physics*, pages 281–300. Academic Press, 1967 (cited on page 164).

[SW78]   Allen J. Schwenk and Robin J. Wilson. Eigenvalues of graphs. In
            Selected Topics in Graph Theory, pages 307–336. Academic Press,
            1978 (cited on page 38).

[Sys73]   Maciej M. Sysło. A new solvable case of the traveling salesman prob-
            lem. *Mathematical Programming*, 4:347–348, 1973 (cited on page 32).

[Sys82]   Maciej M. Sysło. A labeling algorithm to recognize a line digraph
            and output its root graph. *Information Processing Letters*, 15(1):28–
            30, August 1982 (cited on page 32).

[Sze15]   Balázs Szegedy. An information theoretic approach to Sidorenko's
            conjecture, January 2015. arXiv: 1406.6738v3 [math.CO] (cited on
            page 164).

[Täu12]   Hanjo Täubig. The number of walks and degree powers in di-
            rected graphs. Technical report TUM-I123, Department of Com-
            puter Science, Technische Universität München, April 2012 (cited
            on page xiii).

[Täu14]   Hanjo Täubig. Further Results on the Number of Walks in Graphs
            and Weighted Entry Sums of Matrix Powers. Technical report TUM-
            I1412, Department of Computer Science, Technische Universität
            München, July 2014 (cited on page xiii).

[Täu15a]   Hanjo Täubig. *Inequalities for Matrix Powers and the Number of
            Walks in Graphs*. Habilitation thesis, Department of Computer Sci-
            ence, Technische Universität München, 2015 (cited on page xiii).

[Täu15b]   Hanjo Täubig. Inequalities for the Number of Walks in Subdivi-
            sion Graphs. Technical report TUM-I1525, Department of Computer
            Science, Technische Universität München, August 2015 (cited on
            pages xiii, 101).

[Tau60]   Olga Taussky. Problems for solution: 4894. *The American Mathe-
            matical Monthly*, 67(3):294–295, March 1960 (cited on page 53).

[TC10]   Roberto Todeschini and Viviana Consonni. New local vertex invari-
            ants and molecular descriptors based on functions of the vertex de-
            grees. *MATCH: Communications in Mathematical and in Computer
            Chemistry*, 64(2):359–372, 2010 (cited on page 80).

[TM61]   Olga Taussky and Marvin Marcus. An inequality for Hermitian
            matrices (Solution to Problem 4894). *The American Mathematical
            Monthly*, 68(2):185–186, February 1961 (cited on page 54).

[TPR45]   William J. Taylor, Joan M. Pignocco and Frederick D. Rossini.
            Method for calculating the properties of hydrocarbons and its appli-
            cation to the refractive indices, densities, and boiling points of the
            paraffin and monoolefin hydrocarbons. *Journal of Research of the
            National Bureau of Standards*, 34(5):413–434, May 1945 (cited on
            page 28).

[Tri14]   Irene Triantafillou. On the energy of graphs. In *Topics in Mathe-
            matical Analysis and Applications*. Volume 94, Springer Optimiza-
            tion and Its Applications, pages 699–714. Springer, 2014 (cited on
            page 155).

[Tuz93]   Antonín Tuzar. Remark to a problem on 0-1 matrices. *Compositio Mathematica*, 86(1):97–100, 1993 (cited on page 125).

[TW12]   Hanjo Täubig and Jeremias Weihmann. Inequalities for the Number of Walks, the Spectral Radius, and the Energy of Graphs. Technical report TUM-I1211, Department of Computer Science, Technische Universität München, July 2012 (cited on page xiii).

[TW14]   Hanjo Täubig and Jeremias Weihmann. Matrix power inequalities and the number of walks in graphs. *Discrete Applied Mathematics*, 176:122–129, October 2014 (cited on page xiii).

[TWK+13]   Hanjo Täubig, Jeremias Weihmann, Sven Kosub, Raymond Hemmecke and Ernst W. Mayr. Inequalities for the number of walks in graphs. *Algorithmica*, 66(4):804–828, August 2013 (cited on pages xiii, 48, 85).

[Van11]   Piet Van Mieghem. *Graph Spectra for Complex Networks*. Cambridge University Press, 2011 (cited on pages 27, 50, 149, 152, 153).

[Var00]   Richard S. Varga. *Matrix Iterative Analysis*, volume 27 of *Springer Series in Computational Mathematics*. Springer, 2nd edition, 2000 (cited on page 135).

[vDam98]   Edwin R. van Dam. A Cauchy-Khinchin matrix inequality. *Linear Algebra and its Applications*, 280:163–172, September 1998 (cited on page 125).

[VG07]   Damir Vukičević and Ante Graovac. Comparing Zagreb M_1 and M_2 indices for acyclic molecules. *MATCH: Communications in Mathematical and in Computer Chemistry*, 57(3):587–590, 2007 (cited on pages 92, 93).

[VG10]   Damir Vukičević and Marija Gašperov. Bond additive modeling 1. Adriatic indices. *Croatica Chemica Acta*, 83(3):243–260, October 2010 (cited on page 30).

[VGS+10]   Piet Van Mieghem, Xin Ge, Phillip Schumm, Stojan Trajanovski and Huijuan Wang. Spectral graph analysis of modularity and assortativity. *Physical Review E*, 82(5):056113, November 2010 (cited on page 34).

[Vir90]   Ari Virtanen. *Schur-konveksisuudesta ja sen sovellutuksista matriisiepäyhtälöihin*. Licentiate's thesis, University of Tampere, 1990 (cited on pages 127, 129).

[vRW65]   Arnoud C. M. van Rooij and Herbert S. Wilf. The interchange graph of a finite graph. *Acta Mathematica Academiae Scientiarum Hungaricae*, 16(3-4):263–269, September 1965 (cited on page 31).

[VTL79]   Jacobo Valdes, Robert E. Tarjan and Eugene L. Lawler. The recognition of series parallel digraphs. In *Proceedings of the 11th Annual ACM Symposium on Theory of Computing (STOC'79)*, pages 1–12, 1979 (cited on page 32).

[VWG⁺10] Piet Van Mieghem, Huijuan Wang, Xin Ge, Siyu Tang and Fernando A. Kuipers. Influence of assortativity and degree-preserving rewiring on the spectra of networks. *The European Physical Journal B*, 76(4):643–652, August 2010 (cited on page 34).

[Wan08] Hua Wang. Extremal trees with given degree sequence for the Randić index. *Discrete Mathematics*, 308(15):3407–3411, August 2008 (cited on page 89).

[Whi31] Hassler Whitney. Non-separable and planar graphs. *Proceedings of the National Academy of Sciences of the United States of America*, 17(2):125–127, February 1931 (cited on page 31).

[Whi32] Hassler Whitney. Congruent graphs and the connectivity of graphs. *American Journal of Mathematics*, 54(1):150–168, January 1932 (cited on page 31).

[Wie47] Harry Wiener. Structural determination of paraffin boiling points. *Journal of the American Chemical Society*, 69(1):17–20, January 1947 (cited on page 28).

[Wie48] Harry Wiener. Vapor pressure-temperature relationships among the branched paraffin hydrocarbons. *The Journal of Physical Chemistry*, 52(2):425–430, February 1948 (cited on page 28).

[Wil67] Herbert S. Wilf. The eigenvalues of a graph and its chromatic number. *Journal of the London Mathematical Society*, 42:330–332, 1967 (cited on page 151).

[Wil86] Herbert S. Wilf. Spectral bounds for the clique and independence numbers of graphs. *Journal of Combinatorial Theory, Series B*, 40(1):113–117, February 1986 (cited on pages 27, 152).

[Wim85] Harald K. Wimmer. On a weighted mean inequality for nonnegative symmetric matrices. *Linear and Multilinear Algebra*, 17(1):25–27, 1985 (cited on page 61).

[WTPR45] Charles B. Willingham, William J. Taylor, Joan M. Pignocco and Frederick D. Rossini. Vapor pressures and boiling points of some paraffin, alkylcyclopentane, alkylcyclohexane, and alkylbenzene hydrocarbons. *Journal of Research of the National Bureau of Standards*, 35(3):219–244, September 1945 (cited on page 28).

[WW09] Stephan G. Wagner and Hua Wang. On a problem of Ahlswede and Katona. *Studia Scientiarum Mathematicarum Hungarica*, 46(3):423–435, September 2009 (cited on page 70).

[XFS12] Guang–Hui Xu, Kun-Fu Fang and Jian Shen. Bounds on the spectral radii of digraphs in terms of walks. *Applied Mathematics and Computation*, 219(8):3721–3728, December 2012 (cited on pages 148, 153).

[Xu12] Xinli Xu. Relationships between harmonic index and other topological indices. *Applied Mathematical Sciences*, 6(41):2013–2018, 2012 (cited on page 83).

[Yam67] Tetsuro Yamamoto. On the extreme values of the roots of matrices. *Journal of the Mathematical Society of Japan*, 19(2):173–178, April 1967 (cited on page 138).

[YLT04] Aimei Yu, Mei Lu and Feng Tian. On the spectral radius of graphs. *Linear Algebra and its Applications*, 387:41–49, August 2004 (cited on pages 151, 156).

[YLT05] Aimei Yu, Mei Lu and Feng Tian. New upper bounds for the energy of graphs. *MATCH: Communications in Mathematical and in Computer Chemistry*, 53(2):441–448, 2005 (cited on page 156).

[Zho00] Bo Zhou. On the spectral radius of nonnegative matrices. *The Australasian Journal of Combinatorics*, 22:301–306, September 2000 (cited on pages 151, 156).

[Zho04] Bo Zhou. Energy of a graph. *MATCH: Communications in Mathematical and in Computer Chemistry*, 51:111–118, April 2004 (cited on page 156).

[ZL02]   Xiao-Dong Zhang and Jiong-Sheng Li. Spectral radius of nonnegative matrices and digraphs. *Acta Mathematica Sinica, English Series*, 18(2):293–300, April 2002 (cited on pages 144, 145).

[ZL08]   Bo Zhou and Wei Luo. On irregularity of graphs. *Ars Combinatoria*, 88:55–64, July 2008 (cited on page 25).

Index

{0, 1}-matrix, 5

absolute value, 3
adjacency matrix, 6
alternating iterated kernel, 160
alternating power, 107
alternating walk, 108
AM–GM inequality, *see* inequality of arithmetic and geometric means
arithmetic mean, 16, 79–81, 127, 129, 149
average degree, 6, 25, 37, 131, 151

barbell graph, 93
bilinear form, 11
bipartite graph, 41, 86, 108

Cauchy-Schwarz inequality, 13, 50, 108, 109, 124, 130, 138, 155
Chapman–Kolmogorov equations, 9
characteristic function, 162
characteristic vector, 4, 52, 55, 56, 64
Chebyshev's inequality, 14, 71, 74, 105
chemical graph, 85
closed walk, 7, 41, 65, 68, 147
column sum, 8, 107
    weighted, 12
complete bipartite graph, 90
complete graph, 75
complex conjugate, 3, 4
complex matrix, 5
conjugate transpose, 4
conjugate-linear, 11
conversely ordered, 14, 55, 56, 74

de Bruijn graph, 32

degree, 6
    average, 6, 25, 37, 131, 151
    in-degree, 6, 106, 130, 146, 149
    maximum, 37, 146
    minimum, 37, 146
    out-degree, 6, 106, 130, 146, 149
degree variance, 25
density
    of a formal language, 21
    of a graph, 22, 56
deterministic finite automaton, 21
de Bruijn sequence, 32
diagonal matrix, 5, 38, 39
diagonalizable matrix, 38
directed edge, 5
dot product, 4, 11
doubly-stochastic matrix, 8

edge set, 5
eigenvalue, 37, 38
    largest, 135
eigenvector, 38, 61
energy
    graph energy, 154
Eulerian cycle, 6
Eulerian graph, 107

finite-state machine, 20
forest, 87

geometric mean, 16, 79–81, 127, 129
graph, 5
    bipartite, 41, 86, 108
    chemical, 85
    complete, 75

Milton Keynes UK
Ingram Content Group UK Ltd.
UKHW051952071024
449327UK00026B/2279